Design and Performance Evaluation of Pico-Scale Savonius Hydrokinetic Turbines

Roshni

CHAPTER 1

Introduction

1.0 An Overview

Most clean energy comes from renewable sources, which are widely available and free-of-cost. Different types of small hydropower turbines are used for harnessing hydro energy. Hydro energy is harnessed using several kinds of small hydropower turbines. This chapter begins with an introduction to renewable and hydro energy, and its consumption in the world. This is followed by classifications of hydro turbines such as micro, and Pico scale hydro turbines and their importance. Further, Pico-scale hydro turbines is discussed in the next section as Pico-scale Savonius hydrokinetic turbine with its different designs along with its advantages and disadvantages. This is followed by a motivation of the present work. Finally, the chapter concludes by detailing the organization of the thesis.

1.1 Renewables and Hydro energy

Renewables or renewable energy sources such as solar, hydro, wind, etc. presently play a vital role in meeting the energy demands because of zero fuel cost, clean-form and abundance in nature [1,2]. The world's consumption of renewable energy by regions is shown in Fig.1.1 [3]. As water covers 71% of the earth's surface, converting hydro energy to meet energy needs can be both affordable and sustainable. Hydro energy projects are often divided between small and major hydro energy projects, according to the Ministry of New and Renewable Energy, Government of India [4]. In contrast to huge hydro energy systems, small hydro energy systems may essentially be adapted in distant and rural places of the world for the sustainable generation of electricity. Small hydro energy generation capacity ranges from 10 to 30 MW [5]. Small hydro energy systems come in three basic varieties: Mini (100–2000 kW), Micro (5–100 kW), and Pico hydro energy (less than 5 kW) systems [6].

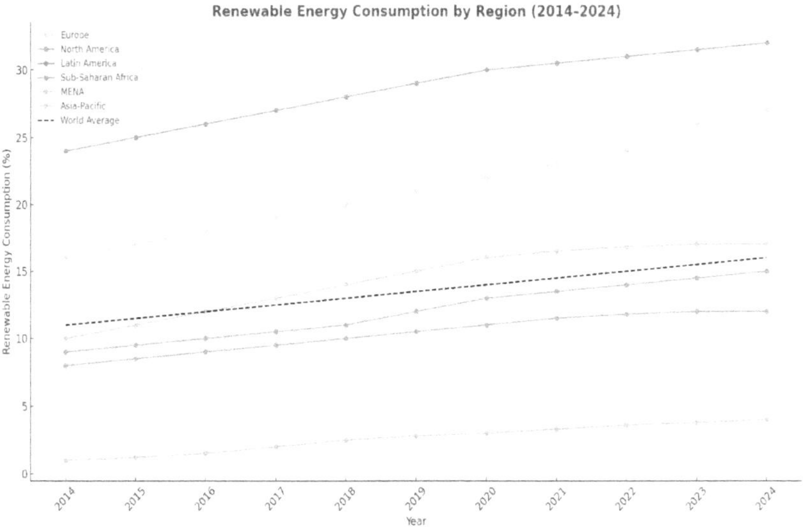

Fig. 1.1 Renewable Energy consumption rate of world by different regions (Bp Statistical Review of World Energy 2022, 2022) [3])

1.2 Micro and Pico Scale Hydro energy and their importance

Micro hydro energy is usually the application of hydroelectric power sized for smaller communities, single families or small enterprise. It is so far the greatest technology available, has a high capacity factor, a high efficiency of 70 to 90%, and can provide energy with reasonable consistency and reliability., i.e., generating power throughout the year; (typically greater than 50% compared with 10% for solar and 30% for wind) [7]. A photograph of a micro hydro turbine (demonstrative type) is shown in Fig.1.2.

Fig.1.2 Micro turbine (Capacity 5-100 kW) [8]

The smallest installation of small hydropower turbines is Pico hydro, which is of capacity below 5 kW as shown in Fig.1.3 (a demonstrative type photograph). Such hydro generators have proven to be useful in small and remote communities that require only a small amount of electricity, such as 50W for a TV or radio and one or two fluorescent lightbulbs. The advantages are as follows:

- Low installation cost.
- Low maintenance.
- Very small infrastructure is required for small scale electricity generation.

Fig.1.3 Pico-scale turbine: capacity less than 5 kW, [9]

1.3 Types of Hydro turbines

Water in motion possesses kinetic energy. Different processes exist for transforming kinetic energy into usable energy. Hydrostatic approach is the conventional way to harness energy from water, which uses the potential head to generate energy. Based on the orientation of the turbine axis with respect to the direction of the water flow, water turbines are of two types: horizontal axis water turbine and vertical axis water turbine. Vertical axis water turbine (VAWT), sometimes referred to as the cross-flow water turbine, revolves around an axis that is perpendicular to the water current [10]. Horizontal axis water turbine (HAWT), on the other hand, has a rotational axis that is parallel to the water current. Hydrokinetic energy conversion is economical and eco-friendly for meeting the energy demands of rural people. Among different hydropower systems, micro hydrokinetic conversion systems are more feasible to extract power from the natural flow of streams and tributaries to provide clean and reliable electricity to remote and rural areas

[6], [11], and [12]. Hydrokinetic energy conversion systems have become popular in the last two decades. Detailed investigations and studies have been done by various researchers on hydrokinetic energy conversion systems. The hydrokinetic energy conversion systems are mainly categorized into water current energy conversion devices. Water current energy conversion system, also called a rotating energy conversion system, is used to harness energy from the flowing stream using the rotating turbine [13]. The turbines used for harnessing hydrokinetic energy come under the micro hydropower systems, where ultra-low head turbines are used for extraction of energy. In-depth assessments of various turbine systems (horizontal axis and vertical axis) for river and tidal applications, along with their classifications and qualitative comparisons, have been done by Khan et al. [14]. Among different hydrokinetic energy systems, river energy conversion systems (RECS) are the most advanced and developed energy systems that are practiced in remote and rural areas. Despite the advancements in river energy conversion systems, some of the pertinent challenges in engineering perspective are mentioned by various researchers. Therefore, progress made in river energy conversion systems with prospects and challenges has been investigated and reviewed in detail, which were covered by [13], [15] and [16]. Humongous research and development have been done by various researchers on harnessing hydropower through various technological advancements. Lago et al. [17] have done a review on the advances and trends in hydrokinetic turbine systems for harnessing hydropower. The current status and emerging designs of hydrokinetic conversion devices technology are discussed in detail in the forms of technical reports, papers, and proceedings [17], [18] and [19].

1.3.1 Horizontal Axis Water Turbine (HAWT)

Horizontal axis water turbine (HAWT) has a rotational axis that is parallel to the flow of free-stream water current. Typically, HAWT has a propeller-like shape with two or three blades, and the lift and drag produced on the blades produces rotating torque. A demonstrative photograph of HAWT is depicted in Fig. 1.4. Solid mooring and buoyant mooring are the two classifications under the HAWT, which may have a straight axis or an inclined axis [20]. Additionally, submerged generators and non-submerged generators are used to categorize buoyant mooring. Different studies on the horizontal axis water turbines are investigated computationally and experimentally. Rajadhyaksha et al. [21] developed a horizontal-axis hydrokinetic turbine system. To achieve the optimal design of the turbine, Prasetyo et al. [22] conducted a simulation study of a horizontal axis water turbine using the flow simulation-Solid works application. McAdam et al. [23], [24], and [25] conducted an extensive investigation on the experimental measurement of hydrodynamic performance and structural loads of a transverse horizontal axis water turbine. Probowoputra et al. [26] conducted a performance analysis of the Savonius horizontal water turbine, considering the effects of turbine stage design. The design and performance prediction of the horizontal-axis water turbine have been examined by Wu et al. [27] using both experimental and computational methods.

Fig.1.4 Horizontal axis water turbine [28]

1.3.2 Vertical-axis water turbines (VAWT)

Hydrokinetic power is generated from the kinetic energy produced by river streams, run-off river, and manmade water channels [29]. In order to utilize energy from zero head and low-current water streams, vertical-axis water turbines (VAWT) are a viable solution. VAWT has its axis perpendicular to the flow of fluid but operates upon the principle of hydrodynamic lift and drag, and its designs have been inherited from the VAWT designs utilized in the wind industry [30]. Due to their simple construction and insensitivity to direction of flow, vertical-axis water turbines are seen to be a practical option for capturing water current energy, especially for low flow speeds [31]. Savonius turbines and Darrieus turbines are two different types of VAHTs. H-type turbines and helical turbines are examples of common Darrieus turbines. As shown in Fig. 1.5(a), an H-type turbine usually refers to a type of vertical axis turbine, where the blades are straight and arranged in an "H" configuration when viewed from the front. Although the H-type turbine has a higher power coefficient, it struggles to start on its own.

Key features:

- Simple construction
- Good for locations with variable flow direction
- Lower efficiency compared to optimized horizontal axis turbines, but cheaper and more rugged

Savonius turbine in Fig.1.5(b), is a type of vertical axis turbine that is simple and mainly used for capturing energy from slow-moving fluids like wind or water. It has a low coefficient of performance but performs well when starting on its own [32], [33].

Applications:

- Small wind energy systems
- Water pumping
- Low-power hydroelectric projects

Advantages:

- Very simple to build
- Starts rotating easily (good starting torque)
- Works well at low speeds
- Accepts flow from any direction (omnidirectional)

Disadvantages:

- Not very efficient compared to lift-based turbines (like Darrieus)
- Slower rotational speed

Theoretically, the Gorlov helical turbine in Fig.1.6(c), has helically twisted blades (like a spiral staircase), attached to a vertical axis. It uses lift forces (not drag) for rotation, just like an airplane wing. Because of the helical shape, it rotates smoothly even in irregular or changing flow (like rivers, tides, or wind). It has a high-power coefficient and strong self-starting performance, however, hydrodynamic performances are greatly influenced by the modification in design parameters.

Applications:

- Tidal energy systems
- River current energy
- Small hydropower without dams

Advantages:

- Very high starting torque (it starts spinning easily)
- Self-starting (unlike older Darrieus turbines)
- Constant torque — no pulsing (smoother rotation)
- Works well in low-speed, multidirectional flows

Disadvantages:

- Slightly more complex manufacturing because of the blade twist
- Needs careful material selection to withstand water forces and corrosion

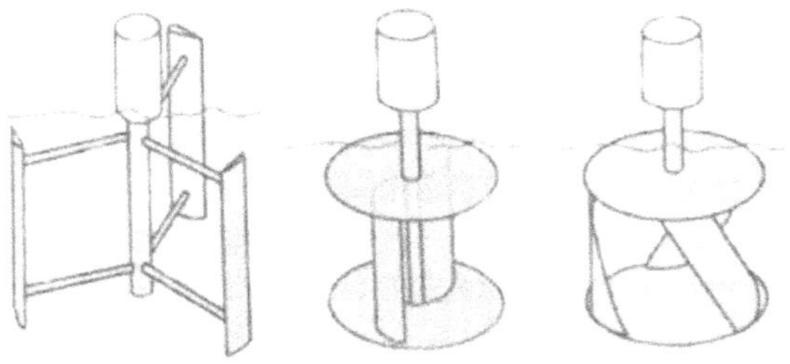

(a) (b) (c)

Fig.1.5 (a) H-Type turbine, (b) Savonius turbine and (c) Gorlov Helical turbine [30]

1.4 Savonius Hydrokinetic Turbine

Savonius Hydrokinetic Turbine (SAHT) is a type of vertical axis water turbine (VAWT). SAHT is simple in design and easy to fabricate at a lower cost as compared to other turbines. Savonius turbine was invented by Finnish engineer Sigurd Johannes Savonius in 1922 [34]. Two patents were filed (the drag-based, S-shaped vertical axis wind turbine): vide number US1697574, published in 1925, and (Improved version of the previous Patent rotor design) vide number US1766765, published in 1928 by Sigurd Johannes Savonius [35], [36]. SAHT technique offers a novel and inventive method for producing electricity in low-velocity flow reserves [37], [38], and [39]. In addition, it is a mechanical device that primarily employs a drag effect mechanism. With SAHT blades, the self-starting technique is flawless. It features a very simple blade profile when compared to more complex airfoil-based designs, such as Darrieus-type turbines [40], [41], and [42]. SAHT blades, which exhibit pressure difference between the two sides of the rotor blades, produce self-spinning for capturing the kinetic energy of moving water [43]. It works like a cup anemometer. Savonius turbines are generally used whenever cost and reliability are more important than efficiency. Large Savonius turbines have been utilized to produce electric power in deep water, which need very low maintenance and generate a good amount of power [44], [45]. Savonius water turbine is good at pumping water, generating electricity, and is not usually connected to power grids. Savonius turbine is also used for driving the extractor fan. With time, the research on conventional Savonius hydrokinetic turbine is modified for harnessing more power from the turbine. The different changes done on SAHT include helical blades [46], twisted blades [47], overlapping [48], multi-staging [49], combined airfoil-straight blade [50], and also bio-inspired blades [51]. The SAHT designs with airfoil-straight bladed Pico-scale turbine is shown in Fig.1.6(a and b).

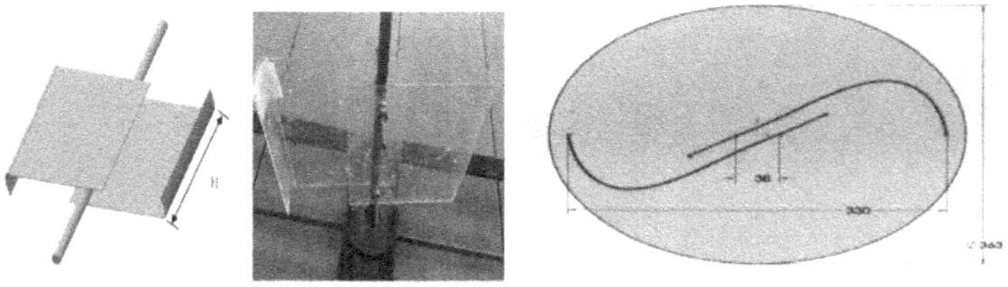

(a) (b)

Fig.1.6 Savonius hydrokinetic turbine with NACA 0018[50] and (b) S1048 airfoil-straight blade [52]

Advantages of airfoil straight-bladed SAHT

- The airfoil curve profile increases the lift, and the straight blade contributes to the drag in the blade segments [50, 52].

- Optimally designed curved shape lessens the generation of negative torque on the turbine, and an extended straight blade provides a longer moment-arm, side-gap, and overlapping flow, thus improving hydrodynamic performance at low flow streams [46].

Disadvantages of airfoil-bladed SAHT

- The fabrication of airfoil-straight bladed Savonius turbine is complex in nature.
- The fabrication requires skilled experts.

1.5 Motivation of the present work

Low flow water-power (less than 5 kW) can be produced in remote areas without grid infrastructure by using runoff from rivers, canals, irrigation channels, etc. However, harnessing water-power from runoff river streams or canals with low free-stream velocities (less than 1 m/s) provides the greatest challenge. Normal hydro-turbines are not practically feasible, since they need a storage reservoir or pond along with head and headwork's. SAHT is a vertical axis turbine that can operate at zero head with flow speed less than 1 m/s, applicable in perennial rivers, channels, and canals. The run-off river locations and irrigation channel for small-scale hydrokinetic power generation are shown in Fig.1.7 (a) and (b). A SAHT type turbine can be used to harvest power from these river tributaries, which primarily meet on flat terrain and have natural or man-made canals with speed less than 1 m/s. Savonius turbines are more common because of their simple designs, self-starting capabilities, low starting torque requirements, omni-directionality, etc. But the major problem is its poor efficiency. Various researchers have increased the efficiency of SAHT using augmentation devices such as deflectors and curtain plates [53], complex blade structures [54], nozzles, etc., at different flow speeds greater than 1 m/s [54]. The motivation behind the present work is the design modification of conventional SAHT for enhancement of performance at flow speed less than 1m/s, and thus to make it applicable for Pico-scale power harvesting.

(a) (b)

Fig.1.7 (a) Run-off-river river channel [55] and (b) Irrigation channel [56]

1.6 Organization of the thesis

The current work has been divided into six chapters -

- Chapter 1 serves as the introduction, providing an overview of the history of renewable energy and hydro energy, Micro and Pico Scale Hydro energy and their importance, types of hydro turbines and their applications, advantages and disadvantages. Airfoil straight-bladed Savonius hydrokinetic turbine is the focus of the study; thus a brief description of them is provided along with a list of their advantages and disadvantages. Finally, the motivation of the present work is also discussed in this chapter.
- Chapter 2 examines the literature assessment on Pico and Micro hydro turbines (Horizontal and vertical axis turbines) studies. Given that the focus of the current dissertation is on the vertical axis Pico-scale Savonius hydrokinetic turbine, a thorough literature study is presented on this topic. The inferences drawn from the literature review are then added to the report. The definition of the problem and the objectives for the current work have also been concurrently described.
- Chapter 3 provides the design, fabrication, and experimentation of the different models of the Pico-scale Savonius hydrokinetic turbines. The methodology to perform the experimental investigations with detailed experimental conditions and details of the different instruments used in the experimental work are also included in this chapter.

- Chapter 4 discusses the computational modeling and optimization of the SAHT models. The methodology of pre-processing, post-processing and Taguchi optimization are briefly explained in this chapter.
- Chapter 5 discusses the analysis of the computational results of the different design models of SAHT, along with their hydrodynamic performances as well as the analysis of Taguchi optimization of optimized model. It provides details of the effect of different arc radii, and blade thickness. The analysis of experimental results by comparing with computational results along with the flow structures around the turbine blades are discussed using in this chapter. This chapter also covers comparison of the obtained results with other published works
- Chapter 6 presents the conclusions of the current work as well as the range of potential future works.

References

[1] Basumatary, M., Biswas, A. and Misra, R.D., 2018. CFD analysis of an innovative combined lift and drag (CLD) based modified Savonius water turbine. Energy Conversion and Management, 174, pp.72-87.

[2] Shahizare, B., Nik-Ghazali, N., Chong, W.T., Tabatabaeikia, S., Izadyar, N. and Esmaeilzadeh, A., 2016. Novel investigation of the different Omni-direction-guide-vane angles effects on the urban vertical axis wind turbine output power via three-dimensional numerical simulation. Energy Conversion and Management, 117, pp.206-217.

[3] bp Statistical Review of World Energy 2022 (71st edition). (2022). https://www.bp.com/content/dam/bp/businesssites/en/global/corporate/pdfs/energy-economics/statistical-review/bp-stats-review-2022-full-report.pdf "Small-Hydro Energy." https://mnre.gov.in/small-hydro.

[4] "Small hydro." https://en.wikipedia.org/wiki/Small_hydro.

[5] "Micro-hydro / Power from the natural flow of streams and small rivers can be harnessed to bring clean, reliable electricity to remote communities." Ashden sustainable solutions, better lives.

[6] http://www.stratosolar.com/2050-world-energy-sankey-diagram.html.

[7] https://www.energy.gov/energysaver/microhydropower-systems

[8] https://free3d.com/3d-model/micro-hydro-turbine-power-generator-4895.html

[9]https://kredl.karnataka.gov.in/new-page/PICO/en

[10] Tian, W., Song, B., Van Zwieten, J.H., Pyakurel, P. and Li, Y., 2016. Numerical simulations of a horizontal axis water turbine designed for underwater mooring platforms. International Journal of naval architecture and ocean engineering, 8(1), pp.73-82.

[11] H. J. Vermaak, K. Kusakana, and S. P. Koko, "Status of micro-hydrokinetic river technology in rural applications: A review of literature," Renew. Sustain. Energy Rev., vol. 29, pp. 625–633, 2014.

[12] L. Wang and R. W. Yeung, "On the performance of a micro-scale Bach-type turbine as predicted by discrete-vortex simulations," Appl. Energy, vol. 183, pp. 823–836, 2016.

[13] M. I. Yuce and A. Muratoglu, "Hydrokinetic energy conversion systems: A technology status review," Renew. Sustain. Energy Rev., vol. 43, pp. 72–82, 2015.

[14] M. J. Khan, G. Bhuyan, M. T. Iqbal, and J. E. Quaicoe, "Hydrokinetic energy conversion systems and assessment of horizontal and vertical axis turbines for river and tidal applications: A technology status review," Appl. Energy, vol. 86, no. 10, pp. 1823–1835, 2009.

[15] M. J. Khan, M. T. Iqbal, and J. E. Quaicoe, "River current energy conversion systems: Progress, prospects and challenges," Renew. Sustain. Energy Rev., vol. 12, no. 8, pp. 2177–2193, 2008.

[16] Güney, M.S. and Kaygusuz, K., 2010. Hydrokinetic energy conversion systems: A technology status review. Renewable and Sustainable Energy Reviews, 14(9), pp.2996-3004.

[17] L. I. Lago, F. L. Ponta, and L. Chen, "Advances and trends in hydrokinetic turbine systems," Energy Sustain. Dev., vol. 14, no. 4, pp. 287–296, 2010.

[18] Ibrahim, W.I., Mohamed, M.R., Ismail, R.M.T.R., Leung, P.K., Xing, W.W. and Shah, A.A., 2021. Hydrokinetic energy harnessing technologies: A review. Energy Reports, 7.

[19] Chaulagain, R.K., Poudel, L. and Maharjan, S., 2023. A review on non-conventional hydropower turbines and their selection for ultra-low-head applications. *Heliyon.*

[20] S. L. L. T. Contreras and O. López, "A review on computational fluid dynamics modeling and simulation of horizontal axis hydrokinetic turbines," J. Brazilian Soc. Mech. Sci. Eng., vol. 6, 2019.

[21] J. T. Simpson and T. N. Us, "(12) United States Patent," vol. 2, no. 12, 2019.

[22] H. Prasetyo, E. P. Budiana, D. Tjahjana, and S. Hadi, "The Simulation Study of Horizontal Axis Water Turbine Using Flow Simulation Solidworks Application," IOP Conf. Ser. Mater. Sci. Eng., vol. 308, no. 1, 2018.

[23] R. A. McAdam, G. T. Houlsby, and M. L. G. Oldfield, "Experimental measurements of the hydrodynamic performance and structural loading of the

Transverse Horizontal Axis Water Turbine: Part 1," Renew. Energy, vol. 59, pp. 105–114, 2013.

[24] R. A. McAdam, G. T. Houlsby, and M. L. G. Oldfield, "Experimental measurements of the hydrodynamic performance and structural loading of the transverse horizontal axis water turbine: Part 2," Renew. Energy, vol. 59, pp. 141–149, 2013.

[25] R. A. McAdam, G. T. Houlsby, and M. L. G. Oldfield, "Experimental measurements of the hydrodynamic performance and structural loading of the Transverse Horizontal Axis Water Turbine: Part 3," Renew. Energy, vol. 59, pp. 82–91, 2013.

[26] D. M. Prabowoputra and J. M. Sohn, "Performance Investigation of the Savonius Horizontal Water Turbine Accounting for Stage Rotor Design," Int. J. Mech. Eng. Robot. Res., vol. 9, no. 2, 2020.

[27] H. N. Wu, L. J. Chen, M. H. Yu, W. Y. Li, and B. F. Chen, "On design and performance prediction of the horizontal-axis water turbine," Ocean Eng., vol. 50, pp. 23–30, 2012.

[28] Verma, D.R. and Katkade, P.S.D., 2018. Horizontal Axis Water Turbine: Generation and Optimization of Green Energy. Int. J. Appl. Eng. Res, 13, p.5767.

[29] M. I. Yuce and A. Muratoglu, "Hydrokinetic energy conversion systems: A technology status review," Renew. Sustain. Energy Rev., vol. 43, pp. 72–82, 2015.

[30] Khan, M.J., Bhuyan, G., Iqbal, M.T. and Quaicoe, J.E., 2009. Hydrokinetic energy conversion systems and assessment of horizontal and vertical axis turbines for river and tidal applications: A technology status review. Applied energy, 86(10), pp.1823-1835.

[31] Kirke, B.K. and Lazauskas, L., 2011. Limitations of fixed pitch Darrieus hydrokinetic turbines and the challenge of variable pitch. Renewable Energy, 36(3), pp.893-897.

[32] Maldar, N.R., Ng, C.Y. and Oguz, E., 2020. A review of the optimization studies for Savonius turbine considering hydrokinetic applications. Energy Conversion and Management, 226, p.113495.

[33] Kumar, A. and Saini, R.P., 2016. Performance parameters of Savonius type hydrokinetic turbine–A Review. Renewable and Sustainable Energy Reviews, 64, pp.289-310.

[34] "Savonius Wind Turbine." https://en.wikipedia.org/wiki/Savonius_wind_turbine.

[35] S. J. Savonius, "Wind Rotor," U.S. Patent 1766765, 1930.

[36] S. J. Savonius, "Rotor adapted to be driven by wind or flowing water," U.S. Patent 1697574, 1929.

[37] Alizadeh, H., Jahangir, M.H. and Ghasempour, R., 2020. CFD-based improvement of Savonius type hydrokinetic turbine using optimized barrier at the low-speed flows. Ocean Engineering, 202, p.107178.

[38] Sinsel, S.R., Riemke, R.L. and Hoffmann, V.H., 2020. Challenges and solution technologies for the integration of variable renewable energy sources—a review. renewable energy, 145, pp.2271-2285.

[39] Sarma, K.C., Biswas, A. and Misra, R.D., 2022. Experimental investigation of a two-bladed double-stage Savonius-akin hydrokinetic turbine at low flow velocity conditions. Renewable Energy, 187, pp.958-973.

[40] Zhang, Y., Kang, C., Zhao, H., and Kim, H.B., 2021. Effects of the deflector plate on performance and flow characteristics of a drag-type hydrokinetic rotor. Ocean Engineering, 238, p.109760.

[41] Sahebzadeh, S., Rezaeiha, A. and Montazeri, H., 2020. Towards optimal layout design of vertical-axis wind-turbine farms: Double rotor arrangements. Energy Conversion and Management, 226, p.113527.

[42] Bazooyar, B. and Darabkhani, H.G., 2020. Design, manufacture and test of a micro-turbine renewable energy combustor. Energy Conversion and Management, 213, p.112782.

[43] Thakur, N., Biswas, A., Kumar, Y. and Basumatary, M., 2019. CFD analysis of performance improvement of the Savonius water turbine by using an impinging jet duct design. Chinese Journal of Chemical Engineering, 27(4), pp.794-801.

[44] Vermaak, H.J., Kusakana, K. and Koko, S.P., 2014. Status of micro-hydrokinetic river technology in rural applications: A review of literature. Renewable and Sustainable Energy Reviews, 29, pp.625-633.

[45] Salleh, M.B., Kamaruddin, N.M. and Mohamed-Kassim, Z., 2019. Savonius hydrokinetic turbines for a sustainable river-based energy extraction: A review of the technology and potential applications in Malaysia. Sustainable energy technologies and assessments, 36, p.100554.

[46] Kamal, M.M. and Saini, R.P., 2022. A numerical investigation on the influence of savonius blade helicity on the performance characteristics of hybrid cross-flow hydrokinetic turbine. Renewable Energy, 190, pp.788-804.

[47] Hassan, M., Iqbal, T., Khan, N., Hinchey, M. and Masek, V., 2010. CFD analysis of a twisted Savonius turbine. *Proc. of NECEC.*

[48] Patel, V., Bhat, G., Eldho, T.I. and Prabhu, S.V., 2017. Influence of overlap ratio and aspect ratio on the performance of Savonius hydrokinetic turbine. *International Journal of Energy Research*, *41*(6), pp.829-844.

[49] Kumar, A., Saini, R.P., Saini, G. and Dwivedi, G., 2020. Effect of number of stages on the performance characteristics of modified Savonius hydrokinetic turbine. Ocean Engineering, 217, p.108090.

[50] Basumatary, M., Biswas, A. and Misra, R.D., 2021. Experimental verification of improved performance of Savonius turbine with a combined lift and drag based blade profile for ultra-low head river application. Sustainable Energy Technologies and Assessments, 44, p.100999.

[51] Hashem, I. and Zhu, B., 2021. Metamodeling-based parametric optimization of a bio- inspired Savonius-type hydrokinetic turbine. *Renewable energy*, *180*, pp.560-576.

[52] Khan, Z.U., Ali, Z. and Uddin, E., 2022. Performance enhancement of vertical axis hydrokinetic turbine using novel blade profile. *Renewable Energy*, *188*, pp.801-818.

[53] Golecha, K., Eldho, T.I. and Prabhu, S.V., 2011. Influence of the deflector plate on the performance of modified Savonius water turbine. Applied Energy, 88(9), pp.3207-3217.

[54] Chen, J., Yang, H.X., Liu, C.P., Lau, C.H. and Lo, M., 2013. A novel vertical axis water turbine for power generation from water pipelines. Energy, 54, pp.184-193.

[55] Run of river. (n.d.).FatStack. https://www.gilkes.com/small-hydropower-solutions/run-of-river-hydropower

[56] Chaudhari, V.N. and Shah, S.P., 2023. Numerical investigation on the performance of an innovative Airfoil-Bladed Savonius Hydrokinetic Turbine (ABSHKT) with deflector. International Journal of Thermofluids, p.100279.

CHAPTER 2

Literature Review

2.0 An Overview

Hydrokinetic energy has the potential to take the role of fossil fuel-based energy without inflicting global warming. The vertical-axis water turbine operates on the same principle as the vertical-axis wind turbine. The relevance of mini, micro, and Pico hydro-power conversion, or small-scale power generation is expanding; thus, researchers are constantly working to find ways to get around the drawbacks of traditional hydro turbines and thus improve their performances. This chapter discusses the literature on turbines in hydro applications and the most current efforts made to enhance the hydrokinetic turbine's performance. In this chapter, the latest research and development works on Pico-scale Savonius hydrokinetic turbines have also been discussed. From the literature reviews, some inferences about the performances and drawbacks of the Pico-scale Savonius hydrokinetic turbines are drawn followed by the genesis of the problem and the problem statement. Finally, the chapter concludes with identifying the broad objectives of the work and methodology for investigation.

2.1 Literature on Hydrokinetic turbines

Water in motion possesses kinetic energy. Different processes exist for transforming kinetic energy into usable energy. Hydrokinetic energy is the energy produced by water when it falls from a higher level to a lower level; this energy is then transformed into electrical energy. The hydrokinetic turbines are categorized as horizontal axis hydro turbines and vertical axis hydro turbines. Horizontal-axis hydrokinetic water turbines (HAHWT) offer a budget-friendly and low-maintenance solution for generating hydropower in channels with low flow velocities [1]. A demonstrative photograph of HAWT is depicted in Fig. 2.1 [2]. With the patent number US8,525,363 B2, Rajadhyaksha et al. [3] have developed a horizontal-axis hydrokinetic turbine system. To achieve the optimal design of the turbine, Prasetyo et al. [4] conducted a simulation study of a horizontal axis water turbine using the flow Simulation-Solid works application. McAdam et al. [5], [6]and [7] conducted an extensive investigation on the experimental measurement of hydrodynamic performance and structural loads of a transverse horizontal axis water turbine. Probowoputra et al.'s [8] performance analysis of the horizontal water turbine took the turbine stage design effect into account. The design and performance prediction of the horizontal-axis water turbine have been examined by Wu et al. [9] using both experimental and computational methods.

Fig.2.1 Demonstrative photograph of HAWT [2]

On the other hand, vertical-axis water turbines (VAWT) are a viable solution to utilize low-velocity currents. The low-velocity current can be obtained from low flow river streams, manmade water channels, and waves [10]. VAWT has its axis perpendicular to the flow of fluid but operate upon the principle of hydrodynamic lift and drag, and its designs have been inherited from the VAWT designs utilized in the wind industry [11]. Due to their simple construction and insensitivity to direction of flow, vertical-axis turbines and are seen to be a practical option for capturing water current energy, especially for low flow speeds [12]. Savonius turbines and Darrieus turbines are two different types of VAWTs. H-type turbines and helical turbines are examples of common Darrieus turbines. As illustrated in Fig. 2.2 (a-c), a helical turbine is an H-type turbine with its blades changed at a specific helix angle. Helical turbines are followed by Savonius turbines and Helical turbines. Although the H-type turbine has a higher power coefficient, it struggles to start on its own. Savonius turbine has a low coefficient of performance but performs well when starting on its own [13], [14]. Theoretically, the helical turbine has a high-power coefficient and strong self-starting performance, however, their hydrodynamic performances are greatly influenced by the modification in design parameters.

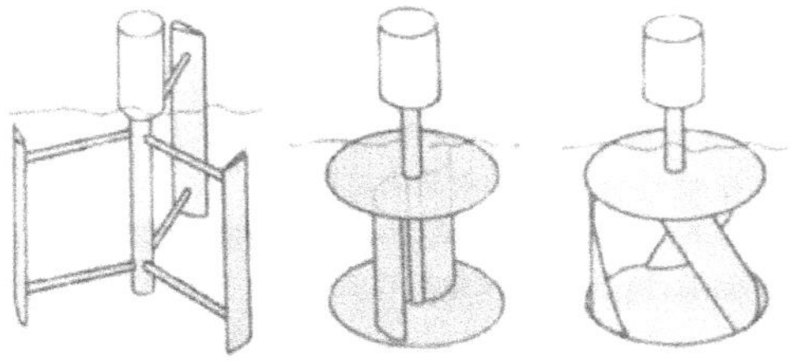
(a) (b) (c)

Fig.2.2 (a) H-Type turbine, (b) Savonius turbine and (c) Helical turbine [11]

2.2 Pico-scale Savonius Hydrokinetic turbine

Savonius hydrokinetic turbines (HKTs) are generally classified as drag-based or lift-based. They are preferred for pico-scale applications due to their simple construction and excellent self-starting capability at low flow velocities, though their power coefficient is relatively low due to significant flow separation and wake losses [11, 15]. Pico hydro systems, generating less than 5 kW, can serve as an economical solution for electrifying isolated rural areas [16]. A recent study on electrification technologies conducted by the World Bank Energy Unit [17] indicates that among the currently available options for off-grid generation, Pico hydro is expected to have the lowest cost (Fig. 1). Maher et al. [18] estimated the costs of Pico hydro schemes in rural Kenya at US 15¢/kWh for 1–2 kW Pico hydro, compared to US $1.09 for a typical 20 Wp solar home system. A Pico-scale Savonius hydrokinetic turbine refers to a very small, low-power water turbine that uses the Savonius rotor design to generate electricity from the kinetic energy of flowing water (like a stream or river). Pico hydro turbine generally refers to systems producing under 5 kW [19], often just a few hundred watts. It has a simple design—typically two or three curved blades (often shaped like an "S" in cross-section). Many researchers are still working on Pico-scale Savonius hydrokinetic turbines to improve their performances by increasing blade numbers [20], modifying blade geometry [21,22] overlapping [23], multi-staging [24], combining airfoil-straight bladed turbine [25], using bio-inspired blades [26], etc. Innovating blade design is one of the prime focuses for improving the Savonius turbine's performance. Research has been carried out with 2 and 3-bladed semi-circular blades and 2 elliptical blades [27]. The 2-bladed semi-circular and 2-bladed elliptical blades have a maximum *Cp* of 0.28 and 0.21 at 0.89 and 0.78 tip-speed ratio (*TSR*) for 0.8m/s. The numerical investigation of the arc shaped parabolic Savonius hydrokinetic turbine was studied, and the outcomes demonstrated that at 0.98 *TSR*, the greatest *Cp* attained was 0.28 [28]. Numerical analysis was conducted over a hybrid hydrokinetic turbine with blades of two semi-circular blades and three S1046 hydrofoil blades at 0.5 m/s flow stream-speed. From the study, *Cp* of 0.21 was attained at *TSR* 0.9 [29]. Keeping some spacing between the stages may help to improve *Cp* at low TSR [30]. The results showed that for 1000 g brake load and 5 mm gap, the maximum *Cp* of 0.0341 was achieved for 0.5 m/s flow stream-speed.

The contributions of the aspect ratio and the overlapping ratio of Savonius hydrokinetic turbines with two semi-circular blades were examined by [23]. According to the findings, the maximum *Cp* was 0.24 for an aspect ratio larger than 1.8 at an overlapping ratio of 0.11. The performance investigation for two-bladed multi-stage Savonius rotor having overlapping was done both experimentally and numerically [31]. According to the results reported, the maximum *Cp* was found to be 0.132 with five stages and the lowest *Cp*=0.081 with two stages for a flow speed of 3m/s. Experimental and computational investigation of three bladed semi-circular Savonius hydrokinetic turbine was performed by [32]. The results revealed that the maximum *Cp* of 0.39 was observed for a flow speed of 0.65m/s at TSR 0.77 and computational *Cp* was 0.402 with a percentage deviation 5.74% with experimental results was obtained. Two twisted blades of Savonius hydrokinetic turbines (SAHTs) with an aspect ratio of 1.58 were investigated computationally at a flow stream-speed of 1m/s. According to the findings, the maximum *Cp* was observed to be 0.35 at 0.9*TSR* for a twist angle of 12.5° [33].

Researchers have recently concentrated on the airfoil-shaped Savonius turbine blade to enhance the hydrodynamic performances. The airfoil-shaped blade profiles are used to generate lift achieving higher efficiency and power coefficients; however, they often suffer from poor self-starting behavior compare to drag based profiles and structural complexity, limiting their suitability for low-speed applications [34]. A numerical study was performed using NACA 7501 airfoil based Savonius turbine with different camber angles [35]. According to the findings, the NACA 7501 with camber angle of 140° has shown maximum *Cp* of 0.26 with 0.9 *TSR* for a low stream-speed 0.4 m/s. Numerical study was performed using NACA 6409 split airfoil. The study reported that maximum *Cp* (0.25) was obtained for a low stream-speed of 0.8m/s at a TSR of 0.9 [36]. Using only lift based airfoil blade is not enough to produce high *Cp* and torque and reduce flow separation under low water speed conditions typically below 0.5m/s. Therefore, to enhance torque, reduce flow separation, and improve overall performance while retaining self-starting characteristics, recent studies incorporated lift-generating airfoil cross-sections into drag-based HKTs [37] to form hybrid lift-drag based Savonius turbines. The literature reported that combinations of different cut sections of airfoils with straight blade segment increase performance more than only airfoil-based blades. A unique blade design for Savonius turbine performance enhancement from low stream-speed may combine cut section of airfoil blade profile with straight blade profile. In this regard, an airfoil S1048 was utilized in traditional Savonius turbine with straight blade section to improve its efficiency. The long straight edge provides larger moment arm and gap flow while the curved segment decreases the effect of negative torque on the rotor. The findings indicate that *Cp* for S1048 based Savonius turbine with straight blade is 0.15 at TSR 1.0 [38]. A novel Savonius water turbine with airfoil-straight segment blades was studied experimentally. The results show that at a low stream-speed of 0.5 m/s, the torque coefficient (T_{coeff}) is 0.538 at 0.418 *TSR* and the greatest *Cp* is 0.268 for *TSR* 0.549 [39]. Another study was conducted using NACA 4415 airfoil to see the effect of leading-edge tubercles on the wake and vortex characteristics. The findings showed that the leading-edge tubercles decreased the wake region, enhanced lift, and minimized the vortex on the suction side of the blades [40]. A recent CFD study was conducted using bioinspired airfoil blade (koi fish) and according to the study, the *Cp* was enhanced to 0.252 at 0.8 m/s stream-speed at a TSR 1.1 [41]. In [42], the conventional Savonius turbine blade profile was slightly modified using airfoil blades with sharp-edge inside and sharp-edge outside. The exterior sharp-edge of the blade gives superior results compared to the interior sharp-edge, with the highest *Cp* of 0.28 and torque coefficients of 0.4, respectively, for *TSR* of 1.

2.3 Inference Drawn from Literature Review

Based on the above comprehensive literature review carried out for the present work, the following inferences have been drawn:

- As the Pico-scale Savonius hydrokinetic turbine is a drag and lift type of power-generating device, it has the inherent capability to self-start at low flow speed, thereby marking its feasibility to operate at low-speed free stream conditions.
- The Pico-scale Savonius hydrokinetic turbine has low efficiency and generates negative torque. Researchers have attempted to improve the performance of this turbine by overlapping, varying aspect ratio, putting multiple blades, multi-staging (two-stage, three-stage and more), etc.

- The helical and twisted blades with different camber angles have been implemented to improve the performance of the SAHT at low flow speed.
- The technology innovations using various airfoils, blade profile modification e.g., applying different airfoil blade profiles of NACA, S-series, straight blade section, etc. have been implemented to improve the SAHT's performance at low flow speeds.

2.4 Genesis of the Problem

People residing in very remote areas are deprived of power. However, they have lots of water reserves of shallow-depth run-off type. Conventional small hydro turbine will not satisfactorily work in such locations due to mounting issues and low volumetric discharge as they need more flow head and speed to spin the turbine blades. Traditional hydroelectric systems often require large dams and reservoirs, leading to high capital costs, long gestation periods, and adverse environmental and social impacts. In contrast, small-scale hydro technologies—specifically run-of-river systems offer an eco-friendlier and cost-effective alternative. Therefore, alternate solution would be to design some Pico scale efficient hydro turbine, which will be able to operate using the free flowing kinetic energy of water of low water current to produce pico-scale power, and it will be also mostly self-starting in nature. Pico-scale Savonius hydrokinetic turbine is more popular due to its simpler design, self-starting, low starting torque requirement, omni-directionality, etc. But the major issue with the such type of Savonius turbine is its poor efficiency and negative torque generation. People have attempted to modify the design of the Savonius hydrokinetic turbine by applying multi-staging of the Savonius turbines (two-stage, three-stage, etc.), increasing number of blades, changing blade shapes, etc. and also have been successful in improving its efficiency. Still, there are ample opportunities to further improve its efficiency low flow speeds less than 1 m/s and low *TSR* conditions. The measures like blade profile modifications, airfoil-straight blade combinations, arc curvature radii of cut out arc sections may require detail investigations for placing the additional part in integration with the turbine and investigating a number of parameters to update the blade profile, thereby involving time and cost.

In this regard putting simple overlapping in its blades and gap between the blades and controlling flow across its height without any additional component may consume less time and effort than elaborate analysis of the augmenters or a number of blade design modification parameters. Although a number of notable achievements have been reported and implemented towards improving the performance of SAHT, there are still scopes for further research to increase the effectiveness of these turbines. Moreover, its development as an efficient low-flow speed scale energy harvester with water as a fluid has not been paid much attention and still, this is at its early stage of development. Because water has special characteristics such as a high density and is anticipated to exert more drag on the turbine blades, the performance of the turbine when operating under the driving fluid (i.e. water) could lead to variations when compared to air being the driving fluid. Therefore, it is viable to conduct a study of the performance of the Savonius turbine with water as the working fluid. When the turbine will operate in field and will be put to load, the implication of the load on the mechanical design of the turbine for its performance under low-flow speed conditions is also required to be properly modelled.

2.5 Scope of present study

Based on the comprehensive literature review and the inferences drawn from it, the research scope for the present work is '*To investigate the performance improvement of Savonius hydrokinetic turbine in Low Stream speeds (≤1m/s) with the help of a combination of design features such as overlapping and gap width ratio, cut out arc sections, combined airfoil-straight blades, arc curvature radii and end plates.*

Therefore, for designing Savonius hydrokinetic turbine system, the cut out curved profile of airfoil is to be attached to straight blades section along with the addition of overlapping and gap width ratio, varying the arc curvature radii of cut out arc section and implementation of end plates. To investigate its performances under low-flow speed conditions, different brake loads are to be put on the turbine.

2.6 Objectives of the present work

To achieve the intended goal, the whole work is divided accordingly, and the objectives for each are accordingly set. These objectives are presented below:

1. **To design airfoil straight bladed based improved Savonius hydrokinetic turbines (SAHTs) for the pico-scale hydro power generation-**
   a. Investigation of different cut-out section of symmetric/asymmetric Airfoils of NACA and S-series.
   b. Modification of the leading edge profile of SAHT using the best airfoil cut-out section

      among NACA and S- series profiles.

c. Designing hybrid blade SAHTs with airfoil cut-out and straight blade sections using
different blade parameters and profiles.

2. To perform computational investigations of the hybrid SAHTs for low flow stream speeds ≤ 1m/s-

a. Investigations of the hybrid SAHT with combination of different parameters such as overlap ratio, gap width ratio, offset and cut out arc sections.
b. Optimization of the performance of the hybrid SAHT using Taguchi design of experiments.
c. Investigations of the optimal design combinations of improved performance of hybrid SAHT.
d. Investigations of the effects of different arc curvatures radii on hybrid SAHT with the optimal design conditions.

3. To perform experimental and computational investigations of the improved hybrid
SAHT for flow stream speed ≤ 1m/s-

a. Experimentally study the effects of different aspect ratios for the hybrid SAHT models.
b. Experimentally study the performance of the hybrid SAHT model with the optimal design parameters.
c. To perform CFD investigations of the experimental SAHT model under the same design and operating conditions.
d. To experimentally study the effect of end plates on the performance of the experimental SAHT model.

2.7 Methodology

The following method is followed for investigating the present study:

Literature review

↓

Inference drawn

↓

Analysis of Research Gap

↓

Design and fabrication of SAHT models

↓

Computational and experimental investigations of the performances of SAHT models in an open channel at various flow speeds and different design conditions.

↓

To obtain the best SAHT model in terms of performance by combining the best design conditions

To obtain the best SAHT model in terms of performance by combining the best design conditions

References

[1] Ferraiuolo, R., Pugliese, F., Álvarez, E.Á., Yosry, A.G., Giugni, M. and Del Giudice, G., 2024. Experimental and numerical investigation of a three-blade horizontal axis hydrokinetic water turbine (HAHWT) in high blockage conditions. *Renewable Energy*, *237*, p.121640.

[2] SOE Professors Receive Grant to Develop New Hydrokinetic Turbines | Rutgers School of Engineering, n.d

[3] J. T. Simpson and T. N. Us, "(12) United States Patent," vol. 2, no. 12, 2019.

[4] H. Prasetyo, E. P. Budiana, D. Tjahjana, and S. Hadi, "The Simulation Study of Horizontal Axis Water Turbine Using Flow Simulation Solidworks Application," IOP Conf. Ser. Mater. Sci. Eng., vol. 308, no. 1, 2018.

[5] R. A. McAdam, G. T. Houlsby, and M. L. G. Oldfield, "Experimental measurements of the hydrodynamic performance and structural loading of the

Transverse Horizontal Axis Water Turbine: Part 1," Renew. Energy, vol. 59, pp. 105–114, 2013.

[6] R. A. McAdam, G. T. Houlsby, and M. L. G. Oldfield, "Experimental measurements of the hydrodynamic performance and structural loading of the transverse horizontal axis water turbine: Part 2," Renew. Energy, vol. 59, pp. 141–149, 2013.

[7] R. A. McAdam, G. T. Houlsby, and M. L. G. Oldfield, "Experimental measurements of the hydrodynamic performance and structural loading of the Transverse Horizontal Axis Water Turbine: Part 3," Renew. Energy, vol. 59, pp. 82–91, 2013.

[8] D. M. Prabowoputra and J. M. Sohn, "Performance Investigation of the Savonius Horizontal Water Turbine Accounting for Stage Rotor Design," Int. J. Mech. Eng. Robot. Res., vol. 9, no. 2, 2020.

[9] H. N. Wu, L. J. Chen, M. H. Yu, W. Y. Li, and B. F. Chen, "On design and performance prediction of the horizontal-axis water turbine," Ocean Eng., vol. 50, pp. 23–30, 2012.

[10] M. I. Yuce and A. Muratoglu, "Hydrokinetic energy conversion systems: A technology status review," Renew. Sustain. Energy Rev., vol. 43, pp. 72–82, 2015.

[11] Khan, M.J., Bhuyan, G., Iqbal, M.T. and Quaicoe, J.E., 2009. Hydrokinetic energy conversion systems and assessment of horizontal and vertical axis turbines for river and tidal applications: A technology status review. Applied energy, 86(10), pp.1823-1835.

[12] Kirke, B.K. and Lazauskas, L., 2011. Limitations of fixed pitch Darrieus hydrokinetic turbines and the challenge of variable pitch. Renewable Energy, 36(3), pp.893-897.

[13] Maldar, N.R., Ng, C.Y. and Oguz, E., 2020. A review of the optimization studies for Savonius turbine considering hydrokinetic applications. Energy Conversion and Management, 226, p.113495.

[14] Kumar, A. and Saini, R.P., 2016. Performance parameters of Savonius type hydrokinetic turbine–A Review. Renewable and Sustainable Energy Reviews, 64, pp.289-310.

[15] Savonius, S.J., 1931. The S-rotor and its applications. *Mechanical engineering*, *53*(5).

[16] Williams, A.A. and Simpson, R., 2009. Pico hydro–Reducing technical risks for rural electrification. *Renewable Energy*, *34*(8), pp.1986-1991.

[17] World Bank, 2006. Technical and economic assessment of off-grid, mini-grid and grid electrification technologies-summary report.

[18] Maher, P., Smith, N.P.A. and Williams, A.S., 2003. Assessment of pico hydro as an option for off-grid electrification in Kenya. *Renewable Energy*, *28*(9), pp.1357-1369.

[19] Thyer, S. and White, T., 2023. Energy recovery in a commercial building using pico-hydropower turbines: An Australian case study. Heliyon, 9(6).

[20] Sarma, K.C., Nath, B., Biswas, A. and Misra, R.D., 2023. Design and performance investigation of a triple blade dual stage Savonius-alike hydrokinetic turbine from low flow stream reserves. Energy Sources, Part A: Recovery, Utilization, and Environmental Effects, 45(4), pp.12099-12117.

[21] Kamal, M.M. and Saini, R.P., 2022. A numerical investigation on the influence of savonius blade helicity on the performance characteristics of hybrid cross-flow hydrokinetic turbine. Renewable Energy, 190, pp.788-804.

[22] Hassan, M., Iqbal, T., Khan, N., Hinchey, M. and Masek, V., 2010. CFD analysis of a twisted Savonius turbine. *Proc. of NECEC*.

[23] Patel, V., Bhat, G., Eldho, T.I. and Prabhu, S.V., 2017. Influence of overlap ratio and aspect ratio on the performance of Savonius hydrokinetic turbine. *International Journal of Energy Research*, *41*(6), pp.829-844.

[24] Kumar, A., Saini, R.P., Saini, G. and Dwivedi, G., 2020. Effect of number of stages on the performance characteristics of modified Savonius hydrokinetic turbine. Ocean Engineering, 217, p.108090.

[25] Basumatary, M., Biswas, A. and Misra, R.D., 2021. Experimental verification of improved performance of Savonius turbine with a combined lift and drag based blade profile for ultra-low head river application. Sustainable Energy Technologies and Assessments, 44, p.100999.

[26] Hashem, I. and Zhu, B., 2021. Metamodeling-based parametric optimization of a bio- inspired Savonius-type hydrokinetic turbine. *Renewable energy*, *180*, pp.560-576.

[27] Talukdar, P.K., Sardar, A., Kulkarni, V. and Saha, U.K., 2018. Parametric analysis of model Savonius hydro-kinetic turbines through experimental and

computational investigations. *Energy Conversion and Management*, *158*, pp.36-49.

[28] Alipour, R., Fardian, F., Koloor, S.S.R. and Petrů, M., 2020. Performance improvement of a new proposed Savonius hydro-kinetic turbine: a numerical investigation. *Energy Reports*, *6*, pp.3051-3066.

[29] Kamal, M.M., Abbas, A., Alam, T., Gupta, N.K. and Khargotra, R., 2023. Hybrid cross- flow hydro-kinetic turbine: Computational analysis for performance characteristics with helical Savonius blade angle of 135°. Results in Engineering, 20, p.101610.

[30] Sarma, K.C., Biswas, A. and Misra, R.D., 2022. Experimental investigation of a two- bladed double-stage Savonius-akin hydro-kinetic turbine at low flow stream-speed conditions. Renewable energy, 187, pp.958-973.

[31] Frikha, S., Driss, Z., Ayadi, E., Masmoudi, Z. and Abid, M.S., 2016. Numerical and experimental characterization of multi-stage Savonius rotors. Energy, 114, pp.382-404.

[32] Sarma, N.K., Biswas, A. and Misra, R.D., 2014. Experimental and computational evaluation of Savonius hydrokinetic turbine for low velocity condition with comparison to Savonius wind turbine at the same input power. *Energy conversion and management*, *83*, pp.88-98.

[33] Kumar, A. and Saini, R.P., 2017. Performance analysis of a Savonius hydrokinetic turbine having twisted blades. *Renewable energy*, *108*, pp.502-522.

[34] Kk Paraschivoiu, I., 2002. *Wind turbine design: with emphasis on Darrieus concept*. Presses inter Polytechnique.

[35] Osama, S., Emam, M., Ookawara, S. and Ahmed, M., 2024. Enhancing the performance of vertical axis hydrokinetic Savonius turbines using a novel cambered hydrofoil profile for rotor blades. Ocean Engineering, 292, p.116561.

[36] Chaudhari, V.N. and Shah, S.P., 2024. Performance enhancement of savonius hydrokinetic turbine using split airfoil blade: A numerical investigation. Renewable Energy, 224, p.120158.

[37] Alom, N. and Saha, U.K., 2019. Evolution and progress in the development of savonius wind turbine rotor blade profiles and shapes. *Journal of Solar Energy Engineering*, *141*(3), p.030801.

[38] Khan, Z.U., Ali, Z. and Uddin, E., 2022. Performance enhancement of vertical axis hydro-kinetic turbine using novel blade profile. Renewable Energy, 188, pp.801-818.

[39] Basumatary, M., Biswas, A. and Misra, R.D., 2021. Experimental verification of improved performance of Savonius-turbine with a combined lift and drag based blade profile for ultra-low head river application. Sustainable Energy Technologies and Assessments, 44, p.100999.

[40] Fan, M., Sun, Z., Yu, R., Dong, X., Li, Z. and Bai, Y., 2023. Effect of leading-edge tubercles on the hydrodynamic characteristics and wake development of tidal turbines. *Journal of Fluids and Structures*, *119*, p.103873.

[41] Hashem, I. and Zhu, B., 2021. Metamodeling-based parametric optimization of a bio-inspired Savonius-type hydrokinetic turbine. Renewable Energy, 180, pp.560-576.

[42] Patel, R. and Patel, V., 2023. Effect of waves on leading edge of modified Savonius rotor blades. Ocean Engineering, 271, p.113445.

CHAPTER

3

Turbine Designs and Experimental Approach

3.0 An Overview

In this chapter, the lift and drag based Pico-scale Savonius hydrokinetic turbine has been designed by modifying the conventional Savonius turbine design. The modification has been done by combining the cut out airfoil profile with the straight blade. For exploration of performance of the proposed turbine, the same has been designed considering all possible design parameters at flow speeds ≤1m/s. Initially, turbine designs are made using CAD software with different design combinations. Finally, for performing experimental investigations, fabrication is done from the best obtained design of computational model.

3.1 Design Specifications of the SAHT models

The Pico-scale Savonius hydrokinetic turbine design consists of two blades having turbine diameter (D) 334.62 mm and blade thickness 1.5 mm, which is obtained by replacing the semicircular curve blade section of conventional Savonius turbine by hybrid blade profile consisting of an airfoil curvature of NACA 63415 at the front-edge, followed by a straight blade section up to the trailing edge. The different design parameters are mentioned below:

3.1.1 Gap width and overlapping

The gap width ratio (ε) is defined as the ratio of the blade gap (g) to the turbine diameter (D_T), which controls the amount of volumetric discharge that can occur in the overlap zone at any flow stream and is represented in Eq. (3.1) [1]:

$$\varepsilon = \frac{g}{D_T} \tag{3.1}$$

The overlapping ratio (φ) is defined as the ratio of overlap distance, e to the diameter of the turbine (D_T) is expressed as Eq.3.2 [2]. Due to overlap of the turbine, flow of water is bounded between the straight blade sections at certain level, which exerts thrust on the straight portion of the blades and thus helps in the rotation of the turbine.

$$\varphi = \frac{e}{D_T} \tag{3.2}$$

3.1.2 Offset ratio

The offset ratio (O_{ff}) is defined as the ratio of the offset (β) to the diameter of the turbine (D_T) is expressed as Eq.3.3. With the implementation of offset in the turbine design, the additional thrust at the straight blade produces momentum to the blade rotation thereby increases the fluid interaction with the blades.

$$O_{ff} = \frac{\beta}{D_T} \tag{3.3}$$

3.1.3 Cut out arc section

Another design element is the cut-out arc section, which is connected to the straight blade section at the leading edge, while the opposite edge remains free as the trailing edge. The cut-out arc section can increase the lift of the turbine blades as same in case of [2] [3], which enhances the overall turbine performances. The cut-out arc section is determined by the corresponding cut out angle (a_1) of the proposed airfoil curvature. The leading-edge geometry was adopted solely to define blade curvature, with a concave orientation intended to improve flow guidance and promote a favourable pressure distribution around the blade. Although the curvature resembles that of a lift-type airfoil leading edge, the present work does not claim classical lift generation on the concave side of the Savonius blade. CFD results confirm that the lift contribution arising from this airfoil-inspired curvature is negligible compared to the dominant drag force. The novelty of this study lies in the adoption of an airfoil-resembling leading-edge curvature as a geometric modification to enhance hydrodynamic interaction and flow attachment, rather than lift infusion. The observed performance improvement is therefore attributed to improved flow guidance and combined drag-dominant behaviour, consistent with terminology used in earlier Savonius turbine studies.

All these design combinations discussed above are shown with the help of schematic as shown in Fig.3.1.

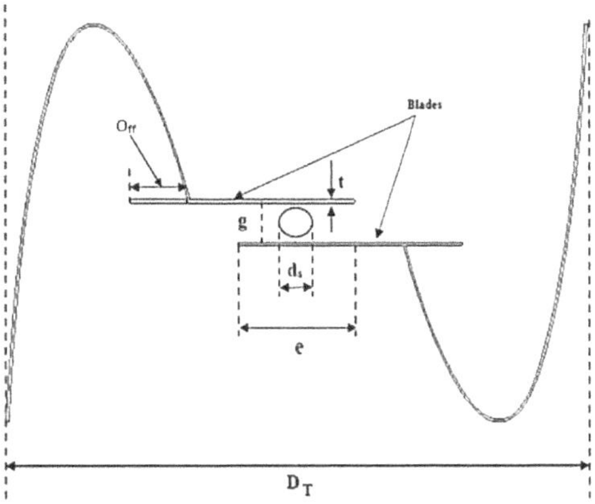

Fig.3.1 Schematic of SAHT with different design combinations

3.2 Design of SAHT models

Different turbine designs are discussed in this section for the Pico-scale Savonius Hydrokinetic turbine. The design includes- (i) Savonius hydrokinetic turbine (SAHT) with gap width but no overlapping ratio and offset (Model-1), ii) SAHT with gap width and overlapping ratio but no offset (Model-2), iii) SAHT with gap width and overlapping ratio and offset (Model-3),

The SAHT(Model-3) is further modified by changing the cut-out arc section of airfoil curvature with combinations of straight blades to form a Model-4. The Model-4 is further modified with three different arc curvature radii (13,15 and 17) mm to form Type-1, Type-2 and Type-3 turbines and vi) Final SAHT (Model-4) is designed with addition of end plates.

The hybrid SAHT is designed using gap width and overlapping ratio, offset, and different cut out arc sections using Pro-E software and the analysis is performed in ANSYS (Fluid flow Fluent). This design modification has been done to improve the performance of the SAHT. The cut out airfoil profile of NACA 63415 and schematic of Model-1,2, 3 and 4 are shown in Fig.3.2(a-e).

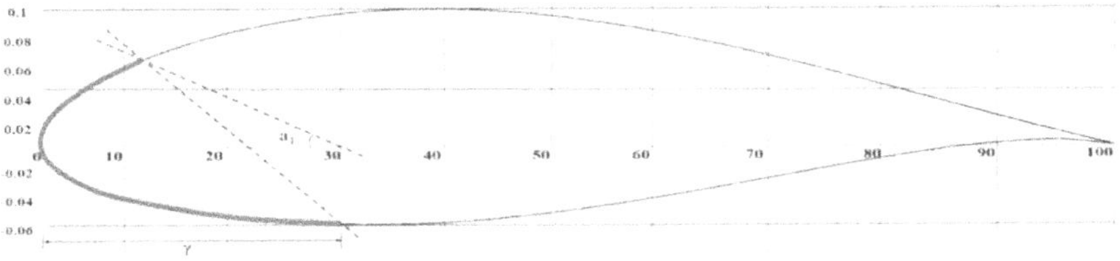

(a)

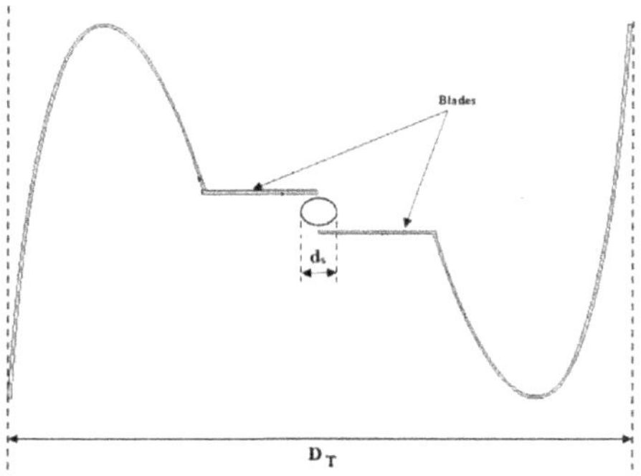

(b)

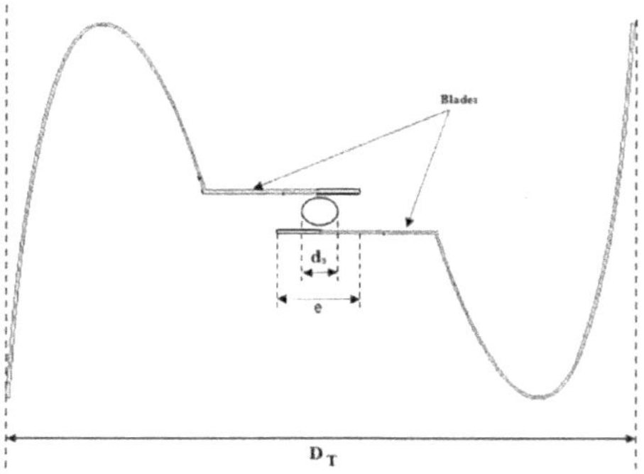

(c)

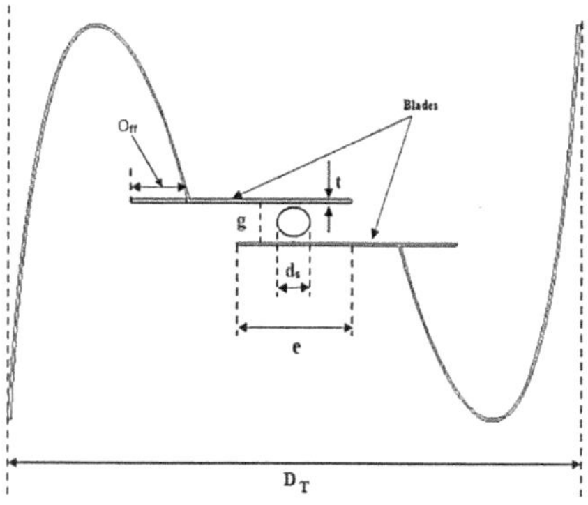

(d)

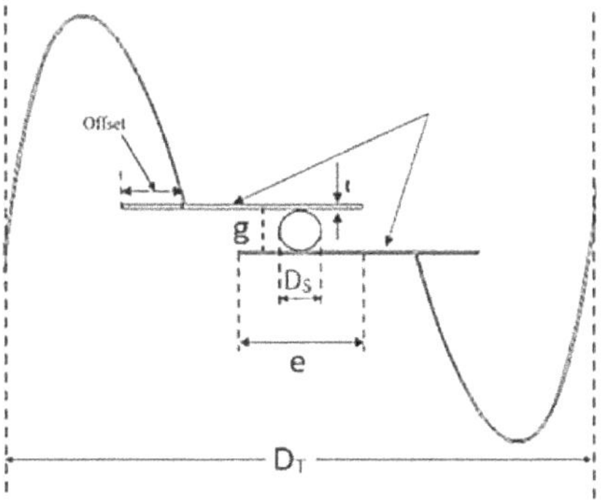

(e)

Fig.3.2. (a)Airfoil profile cut-out from a NACA 63415 blade and Schematic of SHKT (b) (Model-1), (c) (Model-2), (d) Model-3 and (e) Model-4

The model-4 is further replaced with different arc curvature radii(Fig.3.3a) to form a new turbine design viz. (a)Type-1, (b)Type-2 and (c)Type-3 as shown in Fig.3.3(b-d). The CAD model is shown in Fig.3.3

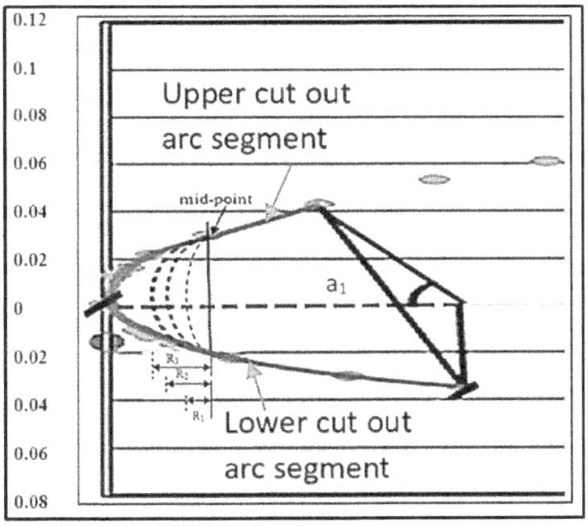

(a)

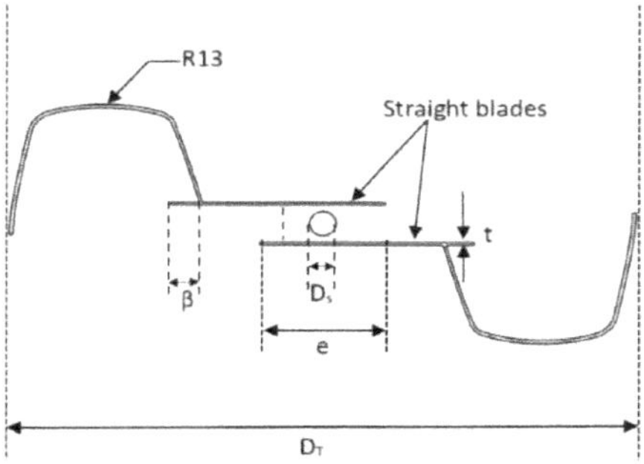
R13
Straight blades
t
β
D_s
e
D_T

(b)

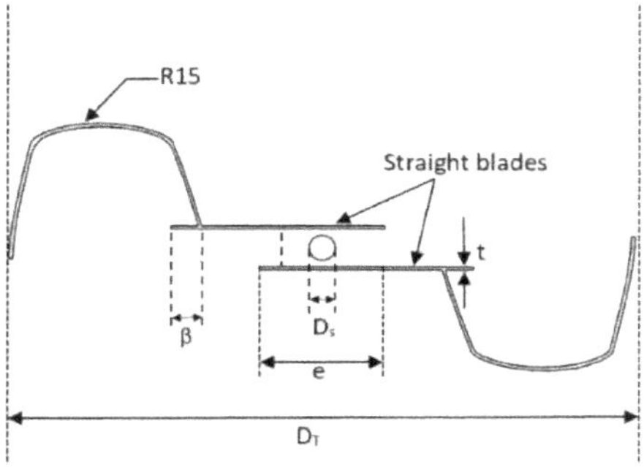
R15
Straight blades
t
β
D_s
e
D_T

(c)

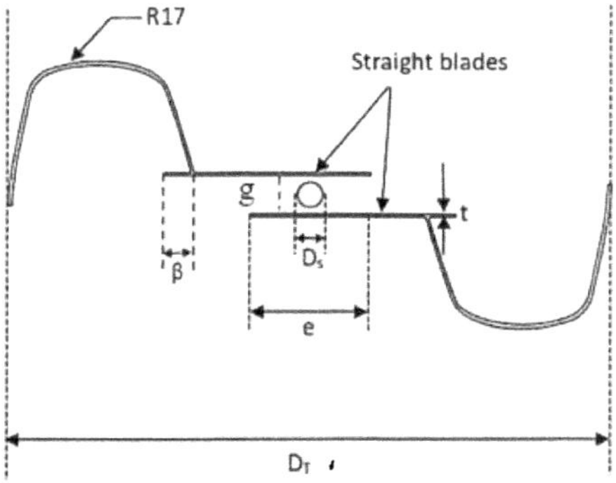
R17
Straight blades
g
t
β
D_s
e
D_T

(d)

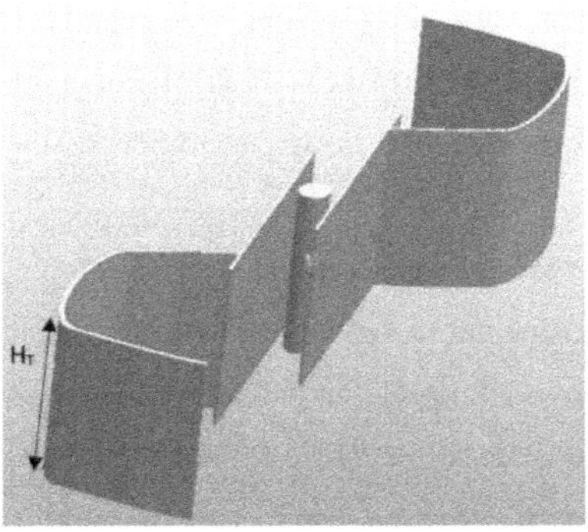

(e)

Fig.3.3 Sketch for (a) Type-1, (b) Type-2 and (c) Type-3 SAHT and (d) 3D CAD turbine model (demonstrative)

The common dimensions of the turbine design for Fig.3.2 and 3.3 are shown in Table 3.1.

Table 3.1 Geometrical parameters for Savonius turbine

Parameters	**Dimensions**
Turbine diameter (D_T)	334.62 mm
Central shaft diameter (d)	14 mm
Turbine height (H)	70 mm
Blade thickness (t)	1.5 mm
Number of blades (n)	2
Overlapping ratio (Φ)	15%,20%,25% and 30%
Gap width ratio (ε)	4.16%,6.14%,6.70% and 7.10%
Cut out arc sections(γ)	11.5%,12%,12.5% and 13%
Cut out arc section angle(a_1)	10°, 27.02°, 28°, 30°, and 33°
Offset(O_{ff})	10mm,15mm,20mm and 25mm
Arc curvature radii (R)	13mm, 15mm and 17mm
End plate diameter	305mm

The overall turbine height for all the models is 0.17 m, the overall turbine diameter for all the models is also same as 0.334 m. As mentioned in Table 3.1, the similar ranges of gap width and overlapping ratio and diameter have been previously studied by [4-7] to determine the hydrodynamic performances of airfoil curvature shaped SAHT. Thus, the present designs follow the standard design condition of the airfoil based SAHT for maximum performance.

3.3 Fabrication of the experimental model

The fabrication of experimental model is very crucial in analyzing the hydrodynamic performances of the Pico-scale SAHT in practical situation. The design parameters in (section 3.2) are implemented in the fabrication of the experimental model. To place the turbine in the water channel, the diameter of the turbine(Model-4) is reduced to 270mm with the scale ratio as 305:1. This is done to minimize the wall side effect on the turbine during the flow. Initially, the blade has been made using 3D printer with the considered dimensions and then final fabrication is done with the aluminium sheets with bending machine following the dimensions of the 3D model. The blades of all the turbine models are made from aluminum sheets with 1.5 mm thickness. All the blades are mounted on a central circular shaft having a diameter of 0.014 m using a nut and bolt arrangement, which facilitates an easy adjustment facility for varying the overlapping and gap width ratio. The turbine shaft for all the models is mounted on a ball bearing which is fixed to a plate frame at the bottom for acquiring support when placed in the water current. The fabricated turbine is shown in Fig.3.4.

Fig.3.4 Fabricated Turbine Model

To perform the experimental investigation, the step-by-step experimental procedure of SAHT model has been discussed as follows:

3.3.1 SAHT with different aspect ratios

The performance of the turbine is greatly influenced by its aspect ratio [8]. A two-bladed Pico-scale Savonius hydrokinetic turbine is mounted on a vertical shaft with different aspect ratio (A_R) of (a)0.22, (b)0.26 and (c) 0.29 has been constructed keeping the other design parameters remain same as discussed in section 3.2. The A_R (0.22 and 0.26) has been selected to keep the turbine blades to always remain submerged in the water channel and A_R=0.29 shows the turbine partially submerged. The experiments are conducted with (A_R) of (a)0.22, (b)0.26 and (c) 0.29 aspect ratios to determine the best aspect ratio.

3.3.2 SAHT with different gap width ratio(ε)

The gap width ratio (ε) generally increases with the distance between the turbine blades, indicating smoother water flow passage between the turbine blades, which can improve turbine performance [1]. The different gap width ratio of 4.16%,6.14%,6.70% and 7.10% are fabricated and investigated experimentally. For brevity, the other gap width ratio on SAHT are not shown.

3.3.3 SAHT with different overlapping ratio(Φ) along with best aspect ratio and gap width ratio

It is known that overlapping increases the performance of Savonius turbines [5, 6, 7]. The overlapping ratio of (15, 20, 25, and 30) % with the best aspect ratio and gap width ratio design combinations are fabricated. The fabricated SAHT model with one of the design combinations of aspect ratio, gap width ratio and overlapping ratio is shown in Fig.3.5. For brevity, the other overlapping ratio design combinations on SAHT are not shown.

Fig.3.5 Top view of the fabricated SAHT with overlap and gap ratio 20%, gap width ratio 6.14% and aspect ratio 0.26

3.3.4 SAHT with aspect ratio, gap width and overlapping ratio along with endplates

A Savonius hydrokinetic turbine (SAHT) is designed with the combination of best aspect ratio, gap width overlapping ratio and offset [9] along with the top and bottom end of the blade joined with end plates as shown in Fig.3.6(a). The scale ratio of the turbine with end plates is 285:1. The top view of the SAHT with end plates is shown in Fig.3.6(b).

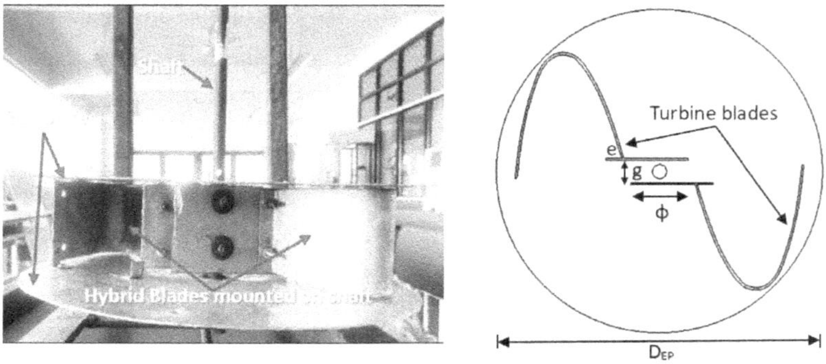

(a) (b)

Fig.3.6 (a) Fabricated Turbine Model, (b) Top view of Turbine

3.4 Experimentation Set-up and Methodology

To test the turbine under fully submerged condition in low stream-speed (< 1 m/s) as found in perennial river streams, run of river streams, etc., the recirculating channel is used with two centrifugal pumps of total capacity of 3.5 Hp. The dimension of the test section in recirculating channel are 4.5 m length, 0.6 m width and 0.395 m height. The maximum achievable water depth in the channel is 0.22 m. The stream-speed and water depth are maintained by operating the inlet and outlet gate lever as shown in Fig.3.7 (a). The fluid from inlet side in recirculating channel is shown in Fig.3.7 (b). Water is pumped from the 2000 L capacity reservoir and re-circulated through the water channel. The rope brake dynamometer set-up is represented in Fig.3.8. The rope brake dynamometer arrangement consists of turbine with vertical shaft, 3 pulleys, rope, load pan and digital weigh scale for measuring loads (g). The arrangement is done in such a way that the rope passes across a pulley from a digital weigh scale, which is then passed around a second pulley at the vertical shaft and after that it returns via another pulley connected to a hook for hanging the weights. The rope is inextensible, and mass of pulley is neglected being very small. Further, all pulleys are properly secured on the structure to not to cause any slip of the rope or any imbalance of the arrangement. The highest water stream-speed of

1 m/s is obtained in water channel. A water current meter (range 0.3-3.5) m/s of an accuracy of ± 0.5% is used to measure the stream-speed. The revolutions of the turbine for a specific load are evaluated using a non-contact digital tachometer that has an accuracy of ± 2%.

(a) (b)

Fig.3.7 (a) Recirculating Tilting channel and (b) Flow of water stream from inlet side in recirculating channel

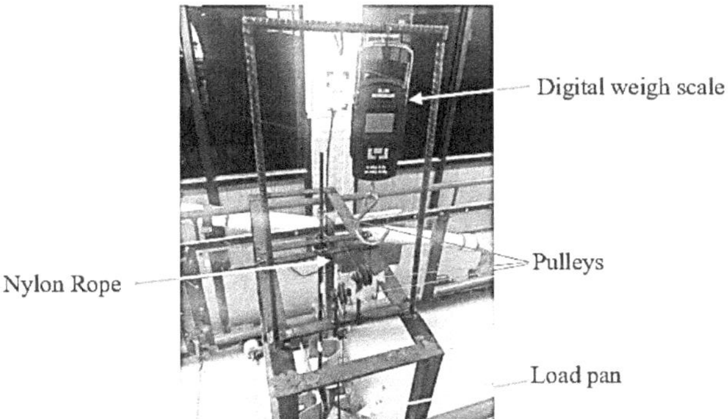

Fig.3.8 Rope-brake dynamometer arrangement [10,11]

The literature [12] has detailed the power loss (maximum) of the turbine models attributed to friction in bearing and the parasitic drag due to nuts and bolts, while the parasitic drag from the nuts and bolts is considered negligible in comparison to the losses caused by bearing friction. The bearing frictional power loss was minimized by opening and washing the seal and by applying proper lubrication to the bearing before operating the model. The maximum bearing power loss is obtained as 0.96% at 1500 g load for a stream-speed 0.8 m/s. Similar levels of bearing frictional power losses were also reported by some previous researchers who worked on such small sized turbines [12], [13]. In the constructed SAHT model with an aspect ratio of 0.26, blockage ratio of the current configuration, after positioning it in the open channel, is determined to be 7.97 % using the Eq.11(Section 3.5). As it is below 15%, this study does not implement any blockage correction [14].

In the considered open channel, a large volume of water is continuously supplied through the inlet section of the channel by two pumps. Although the water flow through the channel is turbulent to a certain extent, the turbulent intensity of it can be controlled. The turbine is to be placed in an appropriate position such that the free stream water current is stable, or in other words, the turbulent intensity of the stream in the upstream direction of the turbine must be within a considerable limit. The turbine is positioned 1.5 m away from the inlet of the channel. After that, preliminary experiments (details presented in the next section) are carried out to confirm flow stability before striking the turbine blades. It is observed that the turbulent intensity of water is within the acceptable limit at this position of the turbine. Since the velocity of the free stream varies for different water depths, the water currents have been measured in different layers of the turbine, counted from the level of the bed. The flow velocities are measuredby the water current meter at three levels of water depth – 0.08 m, 0.12 m, and 0.18 m and average values are considered for the analysis. The free water surface in the channel is 0.03 m above the upper edge of the blades. Keeping parity with the average water flow speeds prevailing in most perennial rivers, the water speeds considered in the present work is (0.4-1.0) m/s.

3.5 Data Reduction

The input-power (P_{in}) can be determined using expression [15, 16] as:

$$P_{in} = 0.5\lambda A V^3 \tag{3.4}$$

Here, λ- density of water (kg/m^3), A- turbine area in m^2 and V- stream-speed in m/s.

The torque (T) of the turbine is determined by the load applied, which is assessed using a rope-brake dynamometer setup. It can be mathematically represented as [17]:

$$T = \frac{(W-S)(R_{pulley} + R_{n.s})}{1000} \tag{3.5}$$

Where, W is the load applied (kgf), S is the residual weigh scale (kgf), $R_{n.s}$ is the radius of the utilized nylon string (m) and R_{pulley} is the pulley radius (m).

Output-power harnessed by the turbine from the incoming flow is:

$$P_{out} = T\omega \tag{3.6}$$

'Cp' known as coefficient of performance as per Eq.3.7 [18,19] is the ratio of power consumed by the turbine (P_{out}) to the available input-power (P_{in}).

$$Cp = \frac{P}{P_{in}} \tag{3.7}$$

Turbine tip-speed ratio (*TSR*) refers to the ratio between speed of turbine blades and free stream-speed of water and represented as in Eq.3.8 [18, 19]:

$$TSR = \frac{\omega R}{V} \tag{3.8}$$

Where, radius and angular velocity of the turbine are represented as 'R' and 'ω', respectively.

Another performance metric considered in this investigation is the torque coefficient (T_{Coeff}), which can be written as follows in Eq. 3.9 [15] as:

$$T_{Coeff} = \frac{T}{0.5\rho A V^2 R} \tag{3.9}$$

Turbulence intensity (T.I.) in the flow at a specific location is assessed as [20], [21] and [22]:

$$\text{T.I.} = \left(\sqrt{\frac{U'^2 + V'^2 + W'^2}{3}}\right) \times \frac{1}{V} \tag{3.10}$$

Where U', V' and W' are the flow velocity in x, y and z directions, respectively.

Blockage ratio (β): Ratio of the turbine frontal area ($H_T \times D_T$) to the cross-sectional area of channel ($H_{ch} \times W_{ch}$) is represented as:

$$\beta = \frac{H_T \times D_T}{H_{Ch} \times W_{Ch}} \tag{3.11}$$

Where, H_T and D_T are the height and diameter of the turbine, H_{Ch} and W_{Ch} are the channel height and width respectively.

3.6 Error analysis

Final outcomes of any experimental analysis always depends on the reliability and correctness of the measured data, which, in turn, depends on human errors, fluctuation in input parameters like voltage fluctuation, and instrumental errors. Due to their nature, the first two categories of errors are of variable type, whereas the instrumental errors are consistent. Human errors can be minimized to a satisfactory level through repeatability tests and data averaging techniques. To handle human error, a total of six numbers of repetitive tests were carried out in the present work to ensure the repeatability of the measurements. Secondly, the measurement errors resulting from voltage fluctuations during experimentations were managed by effective data reductions and smoothing operations to ensure the reliability and correctness of the collected data.

In an experimental setup, in general, several instruments are integrated for measuring different parameters. Each of these instruments, no matter how good the calibration is, has its own degree of deviation from the true value and is generally referred to as the accuracy of the instrument.

3.6.1 Uncertainty Analysis:

In this study, the primary output parameters evaluated are coefficient of performance (Cp) and the torque (T). These parameters are calculated using the relevant measured values as outlined in Eqs. (3.5) to (3.7). Consequently, the instrumental error linked to the uncertainty calculation of Cp is described as [2, 23]-

$$\frac{\delta C_p}{C_p} = \sqrt{\left(\frac{\delta T}{T}\right)^2 + \left(\frac{\delta w}{w}\right)^2 + \left(3\frac{\delta V}{V}\right)^2} \tag{3.12}$$

In a similar way, the following is the expression for the instrumental error related to the turbine's torque calculation:

$$\frac{\delta_T}{T} = \frac{\delta R_{W.S}}{R_{W.S}} \tag{3.13}$$

Where, $R_{w.s}$ is the digital weigh scale reading. The accuracy data pertinent to the instruments utilized in the study are displayed in Table 3.2. Using this information and applying Eqs. (3.12) and (3.13), the total uncertainty linked with the calculated values of Cp and torque (T) has been determined to be ± 3.6% and ± 3.02%, respectively.

Table 3.2. Instrumental accuracy and range

Particulars	**Name of Instrument**		
	Water current meter	**Non-contact Tachometer**	**Digital weigh balance**
Accuracy	±0.5%	± 2%	± 0.29%
Range of values	3.5±0.005 m/s	16±0.02 rpm	0.981±0.0029 Nm

3.7 Preliminary experimentation

3.7.1 Turbulence intensity and flow uniformity

Turbulence intensity (*T.I.*) during the flow must be maintained at a minimal level. Consequently, turbulent intensities are assessed using Eq. (3.10) from the earlier section, measuring stream-speeds at points ranging from 5 cm to 60 cm in 5 cm increments from turbine axis (Fig. 3.9) towards upstream-direction. It has been noted that turbulence intensities progressively decrease as the distance from the turbine increases towards upstream direction, with both the absolute and average values falling below 1%, which is deemed acceptable.

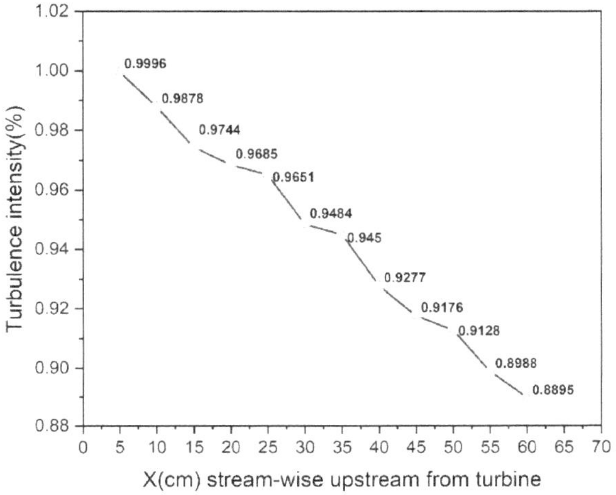

Fig.3.9 T.I. with respect to X (cm) upstream from turbine

Additionally, prior to commencing the main experiments, it is crucial to achieve a nearly uniform stream-speed (m/s) across the entire width and length of the open channel. This step is necessary to minimize the influence of the channel side-walls on the turbine's performance by maintaining a consistent current. To assess the uniformity of the stream-speed in the channel, it is measured at a predetermined free-stream stream-speed (e.g., 0.6 m/s) as illustrated from Fig. 3.10 (a-i). Measurements are taken for 3-different water layers, 1) bottom-layer (8 cm), 2) mid-water layer (12 cm), and 3) top-layer (18 cm) along turbine height, near to left and right of channel's side-walls, and across length of the channel from upstream to the downstream side and resulting stream-speed distributions are examined. The different positions along x-axis are designated to illustrate various cross-stream locations from the left side-wall to the center of the channel (Fig. 3.10 a, c, e), and similarly from the center position to the right sidewall (Fig. 3.10 b, d, f). The stream-speed distributions depicted in Fig. 3.10 (a-f) appear to be nearly uniform. The stream-speed across the length of the channel from the upstream to the downstream (i.e. turbine position) becomes gradually uniform as shown in Fig.3.10(g-i). For the upstream flow with a Reynolds number of Re = 3.58×10^5 (which corresponds to a free stream-speed of 0.6 m/s), the mean water stream-speed measured from both left and right side-wall channel for top-layer (18 cm), mid-layer (12 cm), and bottom-layer (8 cm) are approximately 0.589 m/s as given in Figs. 3.10 (a and b), 0.532 m/s in Figs. 3.10 (c and d), and 0.518 m/s in Figs. 3.10 (e and f), respectively. The percentage error from mean stream-speed for top, middle and bottom layers are 5.17%, 5%, and 7.5%, respectively. Therefore, the mean percentage error for all layers from the mean stream-speed is only 5.89%, which is quite acceptable. The best fitted lines through the points in Fig.3.10 are naturally a bit inclined due to the non-uniformity, which is within acceptable limits. Further, this is worth to mention that similar level of flow uniformity was also noticed in previous research works carried out on the same water channel [24].

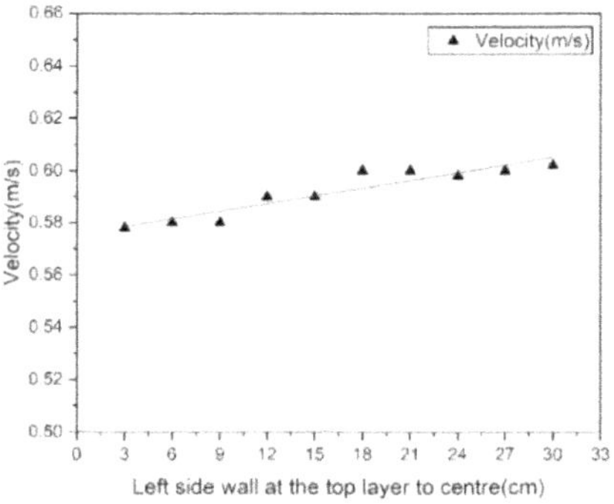

(a)

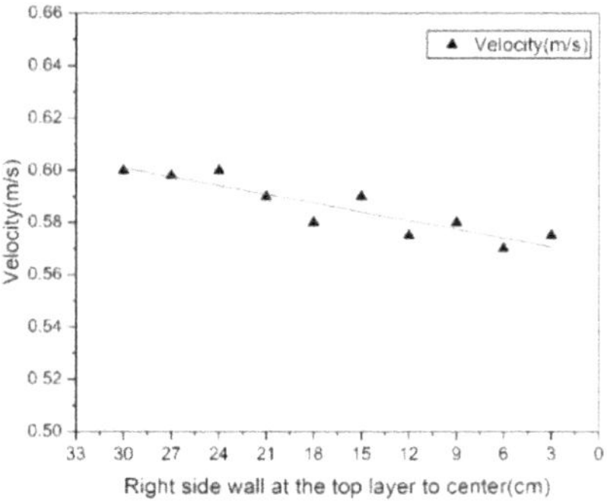

(b)

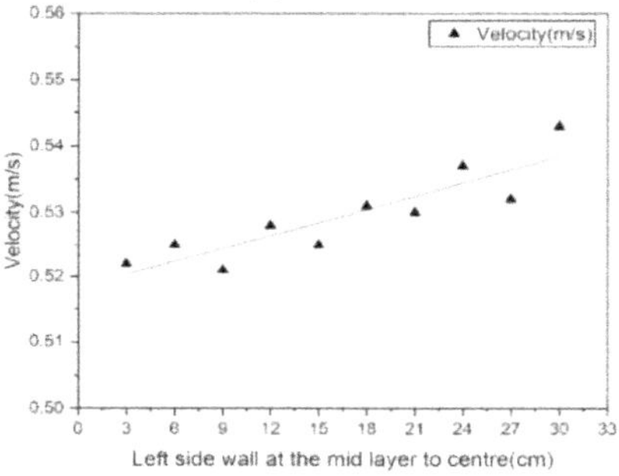

(c)

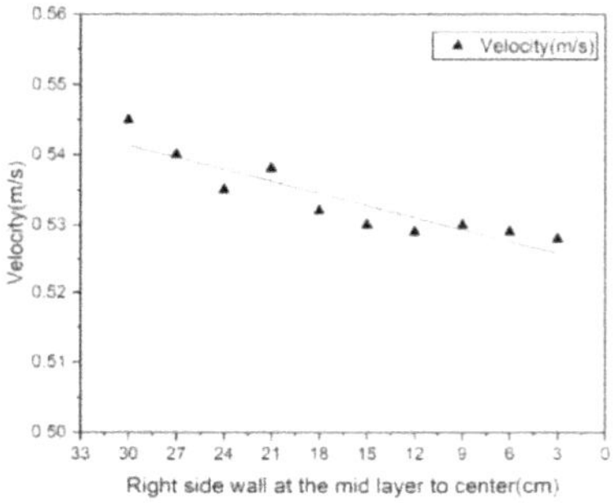

(d)

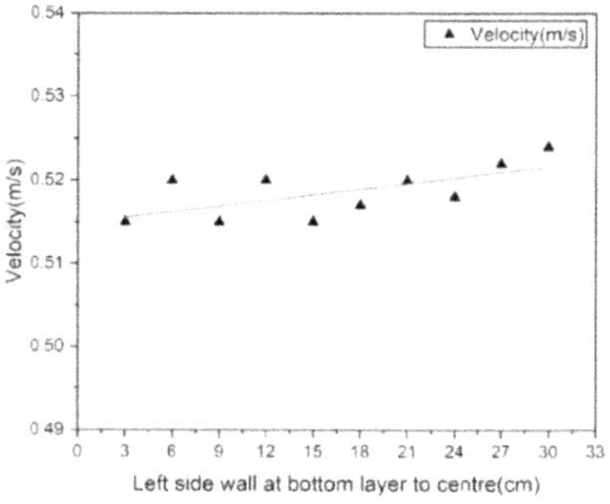

(e)

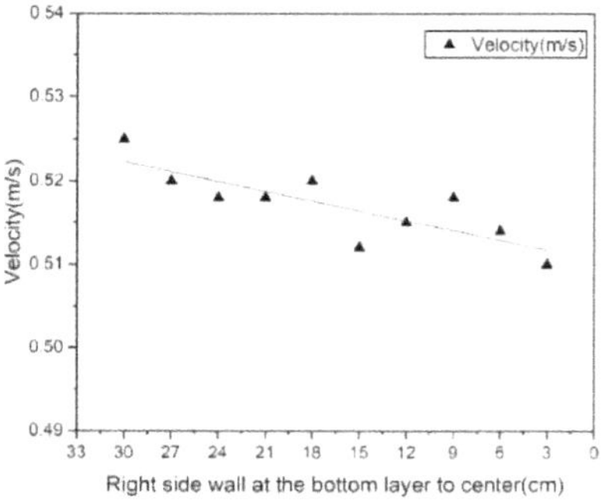

(f)

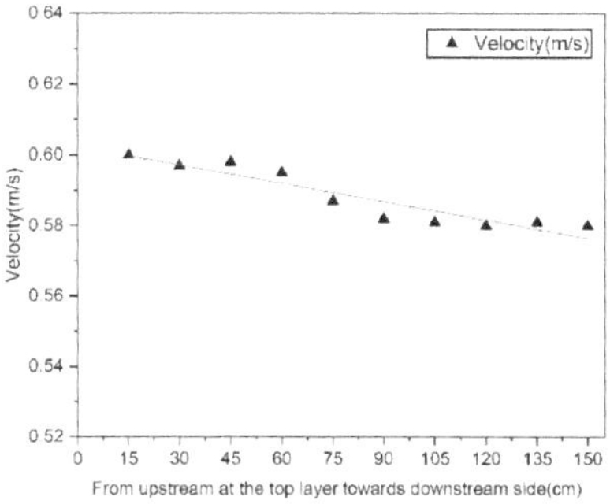

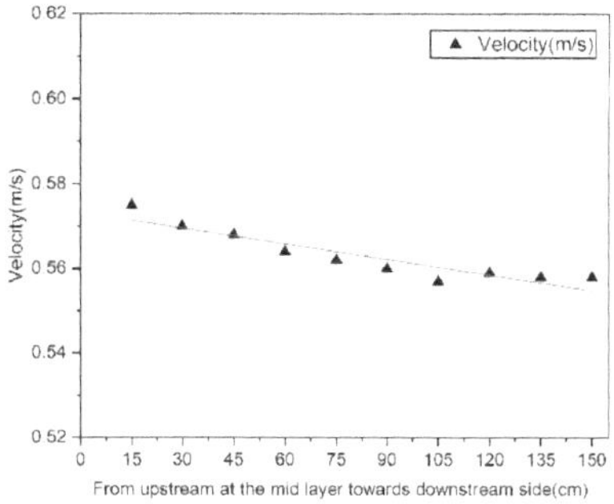

(g) (h)

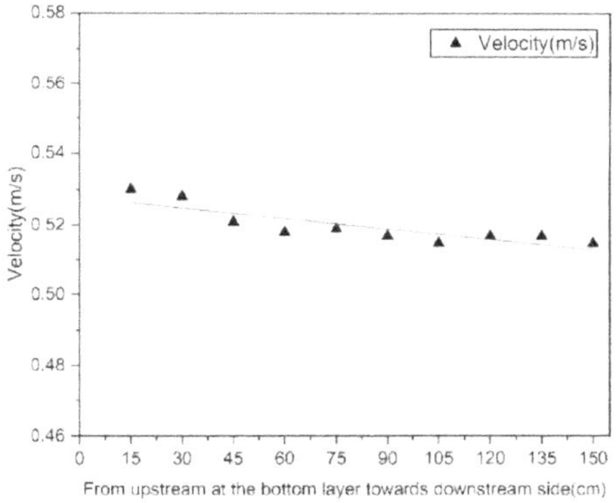

(i)

Fig.3.10 Stream-speed distribution towards centre of channel from (a) left side-wall at the top layer, (b) right side-wall at the top layer, (c) left side-wall at mid-layer (d) right side-wall at mid-layer, (e) left side-wall at bottom layer (f) right side-wall at bottom layer, (g)From upstream at top layer (h) From upstream at mid-layer and (i) From upstream at bottom layer towards downstream of the channel for free-stream-speed 0.6m/s

This Pico-Scale airfoil straight-bladed Savonius water turbine exhibits strong potential for simple manufacturability, low cost, and adaptable deployment in diverse aquatic environments. The straight, airfoil-shaped blades can be fabricated using basic processes such as CNC cutting, 3D printing, or simple jig-based assembly or moulding from materials like plastics, aluminium, or composites, enabling production with minimal specialized tooling—even in community workshops or small fabrication shops. This simplicity translates to lower material and labour costs compared with more complex hydrokinetic turbines, making the technology economically attractive for pico-scale (very small) power generation needs in remote or resource-limited settings. Its inherent design tolerates debris and variable flow directions, and the low operating speed suits slow water currents found in rivers, streams, and irrigation channels, broadening site adaptability without extensive site modification.

References

[1] Basumatary, M., Biswas, A. and Misra, R.D., 2018. CFD analysis of an innovative combined lift and drag (CLD) based modified Savonius water turbine. *Energy Conversion and Management*, *174*, pp.72-87.

[2] Basumatary, M., Biswas, A. and Misra, R.D., 2021. Experimental verification of improved performance of Savonius-turbine with a combined lift and drag based blade profile for ultra-low head river application. Sustainable Energy Technologies and Assessments, 44, p.100999.

[3] Ouro, P., Stoesser, T. and Ramírez, L., 2018. Effect of blade cambering on dynamic stall in view of designing vertical axis turbines. *Journal of Fluids Engineering*, *140*(6), p.061104.

[4] Khan, Z.U., Ali, Z. and Uddin, E., 2022. Performance enhancement of vertical axis hydro-kinetic turbine using novel blade profile. Renewable Energy, 188, pp.801-818.

[5] Sarma, K.C., Biswas, A., Nath, B. and Misra, R.D., 2024. Experimental study of the effect of gap between the stages of a double-stage Savonius Hydrokinetic turbine with end plates and blade overlapping at low water currents. Energy Sources, Part A: Recovery, Utilization, and Environmental Effects, 46(1), pp.10557-10578.

[6] Hassanzadeh, R. and Mohammadnejad, M., 2019. Effects of inward and outward overlap ratios on the two-blade Savonius type of vertical axis wind turbine performance. International Journal of Green Energy, 16(15), pp.1485-1496.

[7] Babu, G.V. and Patel, D.K., 2024. Performance analysis of a Savonius hydrokinetic turbine with aerodynamic blade shape and different overlap ratios: A numerical flow-driven approach. Ocean Engineering, 304, p.117868.

[8] Kadam, A.A. and Patil, S.S., 2013. A review study on Savonius wind rotors for accessing the power performance. *IOSR Journal of Mechanical and Civil Engineering*, *5*, pp.18-24.

[9] Nath, B., Biswas, A., Das, B. and Misra, R.D., 2024. Design, optimization and analysis of a modified Savonius hydro-kinetic turbine (MSHT) with curved winglet and straight blade. Energy Conversion and Management, 314, p.118699.

[10] Kailash, G., Eldho, T.I. and Prabhu, S.V., 2012. Performance study of modified Savonius water turbine with two deflector plates. International Journal of Rotating Machinery, 2012(1), p.679247.

[11] Kamoji, M.A., Kedare, S.B. and Prabhu, S.V., 2009. Experimental

investigations on single stage modified Savonius rotor. Applied Energy, 86(7-8), pp.1064-1073.

[12] Lee, J.H., Lee, Y.T. and Lim, H.C., 2016. Effect of twist angle on the performance of Savonius wind turbine. Renewable Energy, 89, pp.231-244.

[13] Dinh Le, A., Minh, B.D. and Trinh, C.D., 2022. High efficiency energy harvesting using a savonius turbine with multicurve and auxiliary blade. Journal of Fluids Engineering, 144(11), p.111207.

[14] Ramamurthy, A.S., Balachandar, R. and Vo, D.N., 1989. Blockage correction for sharp-edged bluff bodies. Journal of engineering mechanics, 115(7), pp.1569-1576.

[15] Wu, H.N., Chen, L.J., Yu, M.H., Li, W.Y. and Chen, B.F., 2012. On design and performance prediction of the horizontal-axis water turbine. Ocean Engineering, 50, pp.23-30.

[16] Zuo, W., 2005. Introduction of computational fluid dynamics. FAU Erlange-Nümberg, St. Petersburg.

[17] Salleh, M. B., N. M. Kamaruddin, and Z. Mohamed-Kassim. 2020. The effects of deflector longitudinal position and height on the power performance of a conventional Savonius-turbine. Energy Conversion and Management 226:113584.

[18] Kirke, B.K., 2011. Tests on ducted and bare helical and straight blade Darrieus hydro-kinetic turbines. Renewable Energy, 36(11), pp.3013-3022.

[19] Saha, U.K. and Rajkumar, M.J., 2006. On the performance analysis of Savonius rotor with twisted blades. Renewable energy, 31(11), pp.1776-1788.

[20] Talukdar, P.K., Sardar, A., Kulkarni, V. and Saha, U.K., 2018. Parametric analysis of model Savonius hydro-kinetic turbines through experimental and computational investigations. Energy Conversion and Management, 158, pp.36-49.

[21] Alipour, R., Fardian, F., Koloor, S.S.R. and Petrů, M., 2020. Performance improvement of a new proposed Savonius hydro-kinetic turbine: a numerical investigation. Energy Reports, 6, pp.3051-3066.

[22] Kamal, M.M., Abbas, A., Alam, T., Gupta, N.K. and Khargotra, R., 2023. Hybrid cross-flow hydro-kinetic turbine: Computational analysis for performance characteristics with helical Savonius blade angle of 135°. Results in Engineering, 20, p.101610.

[23] Holman, J.P., 2021. Experimental methods for engineers eighth edition.

[24] Shamsuddin, M.S.M. and Kamaruddin, N.M., 2023. Experimental study on the characterization of the self-starting capability of a single and double-stage Savonius-turbine. Results in Engineering, 17, p.100854.

CHAPTER 4

CFD Modelling and Taguchi optimization of the SAHT models

4.0 An Overview

This chapter discusses about the computational study, which consists of Pre-processor and Processor followed by optimization methodology to determine the optimal SAHT design. Initially, different types of airfoils from S, NACA series are selected to determine the best airfoil type. The selection of best airfoil is considered using higher average lift coefficient to drag coefficient. In the pre-processor, the best airfoil is designed using a straight blade section to form a hybrid SAHT models (Model-1 to Model-4). To perform grid and time independency test of SAHT models, the computational domain is created, which is then followed by meshing of the computational domain. Similarly, under Model-4, computational domain of Type-1 to Type-3 turbines are created for each models. For brevity all the computational domains are not shown. Secondly, in the processor, it consists of governing equations, meshing, boundary conditions, solver details, solution methods, grid independency test, time step independency test and validation are discussed in this section. Finally, Taguchi optimization procedure of the SAHT design(Model-3) with different design parameters are discussed in this chapter.

4.1 Pre-Processing

4.1.1 Types of airfoil

Four airfoil-based blade curvatures- two from symmetric airfoils (S1046, S1048) and two from unsymmetrical airfoils (NACA 63415, NACA 63418), are considered in association with the flat blade sections. However, this section presents the NACA 63415 curve profile, which has the higher lift/drag ratio (C_L/C_D=52.7) as per literature [1]. The average value of C_L/C_D for unsymmetrical airfoil profile have been taken to identify the better airfoil in the present study and it has been observed that NACA 63415 has higher C_L/C_D ratio. The higher C_L/C_D ratio provides better performances in airfoil bladed turbine [2]. Furthermore, thin and symmetric airfoils are generally excluded because they neither enhance self-starting capability nor improve net torque production and structurally weaker and more susceptible to deformation under hydrodynamic loading, particularly in water turbines. NACA 63415 in [3] exhibits superior hydrodynamic performance in terms of torque among the considered profiles. Similar design methods as discussed below are adopted for the other three airfoil-based blade curvatures, although not repeated for brevity.

4.1.2 Computational modelling, and domain

The seven models (Model-1 to Model 4) and Type-1 to Type-3 are modelled using Pro-E software and analysis are performed using ANSYS (Fluent) software as per the dimensions discussed in the previous chapter 3. The turbine is enclosed by a rectangle shape and circular shape limits within a computational domain where grids are established. The rectangular limit dimension is 7DT $\times 10D_T$ [4], and circular limit is $3D_T$ [5]. A circular enclosure is established around the turbine area to examine the mesh intricacies and evaluate the close meshing effect over turbine performance. Rectangular enclosures are constructed and held at a considerable distance from the turbine to ensure that the borders do not influence the results, thus allowing for an accurate analysis of the fluid flow through the turbine. Figure 4.1(a) and (b) (demonstrative) displays both rotating and stationary zones of the two different turbine models.

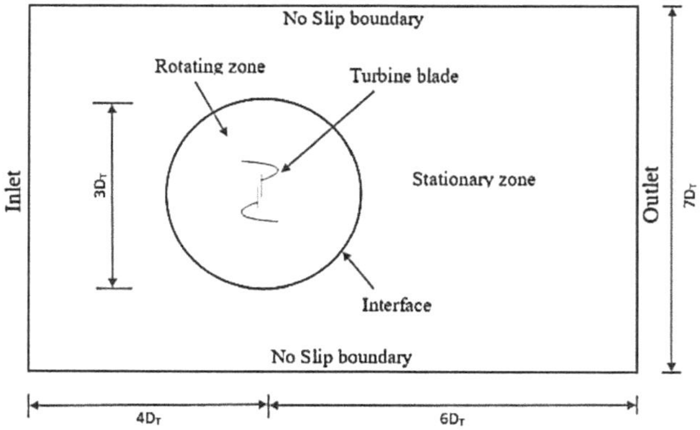

(a)

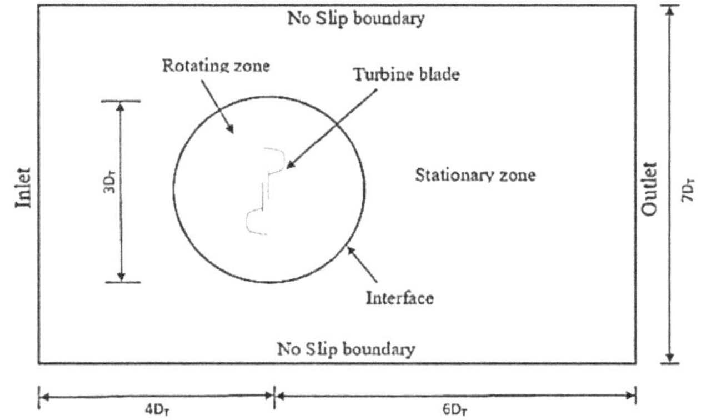

(b)

Fig.4.1 Computational domain of (a)Model-3and (b) Model-4 (demonstrative)

4.1.3 Governing Equations

Flow surrounding the turbine is assumed to be turbulent, unsteady, and incompressible, and the governing equations shown below are controlled by the incompressible Navier-Stokes equation [6, 7].

$$\frac{\partial U_i}{\partial x_i}=0$$

(4.1)

$$\rho(\frac{\partial U_i}{\partial t} + \frac{\partial}{\partial x_j}\left(U_i U_j + \overline{u_{i'} u_{j'}}\right) = -\frac{\partial P}{\partial x_i} + \frac{\partial}{\partial x_j}\left(\mu \frac{\partial \overline{U_{i'}}}{x_j}\right)$$

(4.2)

Where, U_i –speed along 'i [th]' direction, and u' - fluctuation speed.

ρ-density, P-pressure *and* μ-dynamic viscosity, while '*i*' and '*j*' stand for indices along x- and y-direction. The SST(Shear Stress Transport) k-ω turbulent equations were utilized to represent the Reynolds stress term $\rho\overline{u_{i'} u_{j'}}$ in the momentum equation. The eddy viscosity model acts as the basis for the SST k-ω model, which was developed by [8]. SST k-ω model is more accurate and reliable for complicated flow problems such as transonic shock waves, curve profiles, and adverse-pressure-gradient flows. Also, such model performs well in both far-wall and near-wall zones. Below are the formulas for the specific dissipation rate and turbulent kinetic energy of this model [9].

$$\frac{\partial(\rho k)}{\partial t} + \frac{\partial}{\partial x_i}(\rho k u_i) = \frac{\partial y}{\partial x_j}\left(\gamma_k \frac{\partial k}{\partial x_j}\right) + T_k - Y_k$$

(4.3)

$$\frac{\partial(\rho\omega)}{\partial t}+\frac{\partial}{\partial x_i}(\rho\omega u_i)=\frac{\partial y}{\partial x_j}\left(\gamma_\omega\frac{\partial\omega}{\partial x_j}\right)+T_\omega-Y_\omega+d_\omega$$

(4.4)

Where γ_k and γ_ω - effective diffusivity of ω and k.

T_k represents turbulent-kinetic-energy (TKE), and T_ω is production of ω.

Y_k and Y_ω denotes dissipation of k and ω, and d_ω - trans diffusion term.

4.1.4 Meshing procedure

4.1.4.1 Meshing of SAHT models

Three different mesh sizes with different numbers of elements, including coarse, medium, and fine mesh, have been investigated to see the variation of torque. The coarse, medium, and fine meshing obtained as 119255, 131284, and 149528, respectively. The medium meshing of computational domain and meshing of leading edge of cut out section of curve airfoil profile is applied to the turbine model(Model-1), as shown in Fig.4.2 (a) and (b). The meshing of other models are not repeated for brevity. The face sizing of 0.006-0.02 mm and 0.01-0.020 mm for both the inner fluid and outer fluid zones are applied to find the optimal mesh. The number of nodes and elements for the optimal mesh sizes are 52836 and 51870, respectively. The maximum skewness ratio is 0.72, and the average orthogonal quality is 0.44 [10]. To model the performance of the modified turbine, various water stream speeds of 0.3, 0.4, 0.5, and 0.6 m/s are taken into consideration at the computational domain's inlet.

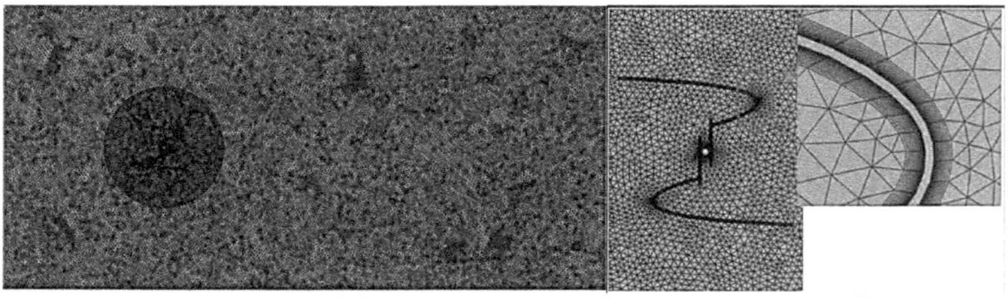

(a) (b)

Fig.4.2 (a) Schematic of meshing and (b) Meshing of leading edge of cut out section of curve airfoil

4.1.4.2 Boundary conditions

Table 4.1 contains a list of boundary conditions. Transient CFD simulation is applied to the suggested turbine types.

Table 4.1 Boundary conditions of HSHT

Region	Type
Left side wall	Flow Inlet
Right Side wall	Pressure-outlet
Top and bottom wall	No-slip condition
Turbine Blade	Rotating wall
Interface	Sliding-mesh
Rotating	Internal zone
Fixed	External zone

4.1.4.3 Solver method and solution control

The computational conditions and the properties of fluid used for the present turbine are as follows-

- Model of computation: SST k-ω
- Solution methods: Momentum-second order upwind; Pressure-second order, Turbulent Kinetic Energy (K.E.), and specific dissipation-second order upwind
- Pressure velocity coupling: SIMPLE (Semi-Implicit Method for Pressure Linked Equations) scheme
- Solution for convergence: Absolute and residual value (0.00001)
- Water density: 998.4 kg/m3
- Water viscosity: 0.001003 kg/m-1 s-1
- Flow stream-speed: 0.3-0.6 ms-1

Transient conditions are established to assess the altered turbine performance.

4.1.4.4 Grid independency test of Model-1

The grid independency test is performed by changing the nodes and elements until an optimal quality of the mesh is achieved. The idea behind adopting varied element sizes for the interior and exterior zones is to improve the quality of the mesh close to the turbine zone; the interior fluid zone element size is considered smaller than the exterior fluid zone. The torque at coarse, medium, and fine meshing for NACA 63415 turbine model is shown in Fig.4.3. It is found that the torque value is low at coarse mesh, and it increases in medium mesh, and then it becomes constant at fine meshing. There is very little variation in torque value between medium and fine mesh sizing. So, medium mesh is considered for further simulation as the solution becomes smooth and fast in medium mesh size. To evaluate the quality of the mesh, wall y+ is computed; changes in y+ for blade 1 and blade 2 are shown in Fig.4.4. It has been noticed that the average y-plus is less than 1.5, which is regarded as within the permitted range for k-ω models [4] [11]. For brevity, the grid independency tests for other models are not repeated.

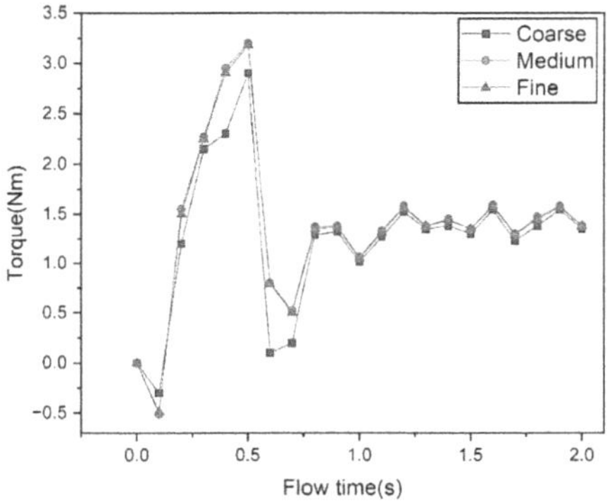

Fig.4.3 Torque variation at coarse, medium and fine mesh

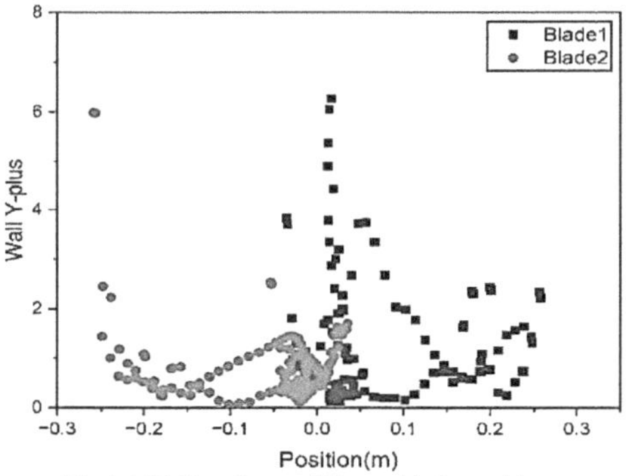

Fig.4.4 Wall y-plus at different blade positions

4.1.4.5 Time-step size independency test of Model-1

The time-step is yet another element that affects the precision and effectiveness of the CFD analyses [7] [12]. Three distinct time-step-sizes—$t_1 = 0.001$ s, $t_2 = 0.0015$ s, and $t_3 = 0.0020$ s have been used in the time-step-size independency studies at constant tip speed ratio value of 0.7, overlapping percentage 20%, side clearance percentage 6.14% and offset 15 mm. Similar time step sizes were also adopted in previous research works [13] [14]. However, these time-step sizes are calculated using the ratio of angular rotation in radian to the circular frequency of the turbine in radian per second. Further, time step size below 1s i.e. from 0.001s onward is chosen to see the effect of flow around the blades in the milliseconds of time. This can provide more accurate findings along with the ability to capture unsteady flow turbulence [7]. According to Fig.4.5, at a time-step size of 0.0015 s, *Cp* is almost close to the time step size of 0.001 and 0.002. In fact, there is marginal difference of values between these time step sizes. Whereas, if time step lower than 0.0015s is chosen, then the simulations consume more time to converge. Hence, to expedite the simulation results, time-step size of 0.0015 will be computationally more efficient, and hence this time-step size is considered for further numerical analysis. For brevity, the time independency tests for other models are not repeated.

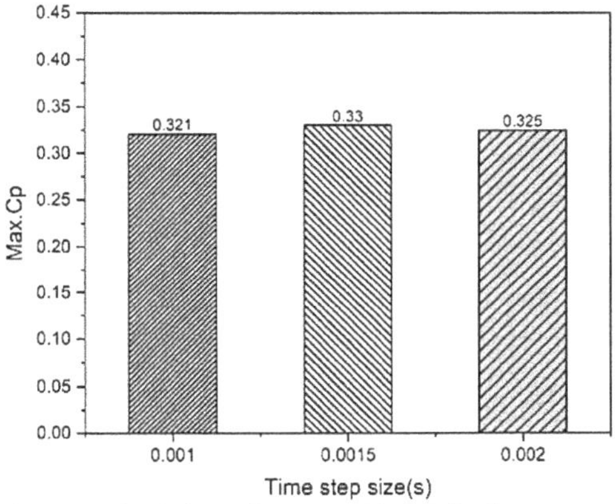

Fig.4.5 Time step size independency at *TSR*=0.7 for the present turbine model

4.1.4.6 Grid and Time independence study for Type-2 turbine

The grid independence study of Type-2 under Model-4 turbine is conducted by adjusting the mesh's elements and nodes until it reaches the ideal quality. The simulations are performed at 0.7 *TSR*. The number of elements in the mesh study is varied by changing the face size of both fixed and rotating domains for coarse, medium, and fine meshing. For coarse, medium, and fine meshing, the number of elements is (138575, 164071, and 175852), (178086, 187131, and 203627) and (189664, 225090, and 236586), which are used in simulation to see the variation of torque value, and from the simulations, the average torque in all the cases are determined, and the errors are calculated. The error percentages obtained for coarse, medium, and fine meshing are 0.87%, 0.14%, and 0.44%, as shown in Table 4.2. It can be seen from Table 4.2 that the average torque value for medium meshing is close to fine meshing, and the relative error (%) for all the mesh sizes is less than 1%, which is quite acceptable [15]. Thus, as the computational process becomes smooth and quick at medium mesh sizes and the average torque is almost the same as that of fine mesh (Table 4.2). Therefore, medium mesh is taken into consideration. Figure 4.6 displays y+ plots for turbine blades to assess the mesh quality. The average y+ is observed to be below 2, which lies in the acceptable range for the k-ω model [4,16]. The average skewness is 0.13, with the mean orthogonal quality as 0.42 [10]. For brevity, the grid independency tests for other types are not repeated.

Another factor influencing the precision and effectiveness of the CFD analysis is the time- step [7, 12]. Time steps are determined by the division of the angular rotation (rad) and circular frequency of the turbine (rad/s). For time step independency investigations, three-time steps, 0.0012 s, 0.0015 s, and 0.0018 s of medium meshing, were employed at a *TSR* 0.7 with T_{avg} values as shown in Fig. 4.7. These similar time steps are also applied in existing studies [13, 14]. A time step size of less than 1 sec is selected to observe the impact on flow surrounding the blades in milliseconds. This has the potential to capture unstable flow turbulence and yield more accurate results [7]. There is little variation between these time steps, and if the time step below 0.0015s is chosen, the simulation will consume more time and become slow to converge. Therefore, time step of 0.0015 is taken into consideration for further numerical analysis since it will be computationally more efficient and speed up the simulation results. For brevity, the time independency tests of Type-1 and Type-3 are not repeated.

Table 4.2 Mesh independence study

Meshing	T_{avg}	Relative Error (%)
Coarse	0.683	0.87
Medium	0.687	0.14
Fine	0.686	0.44

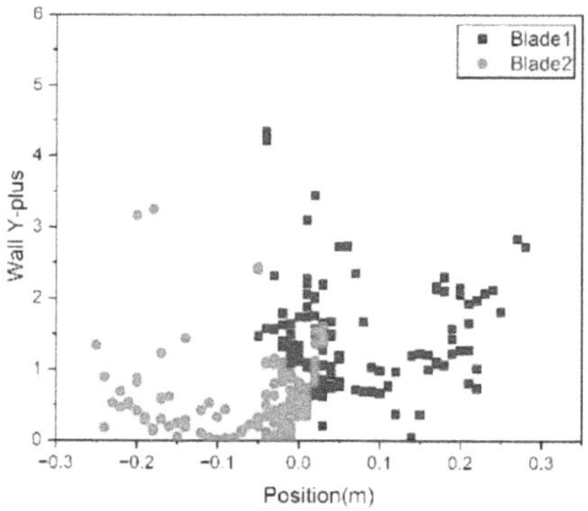

Fig.4.6 Value of wall y+ for two blades positions

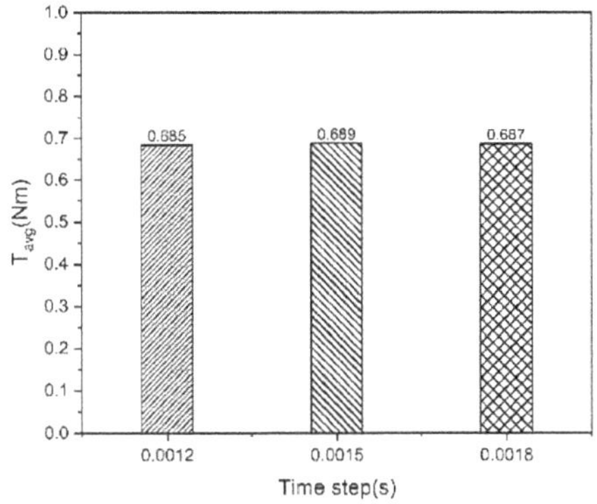

Fig.4.7 Average torque with respect to time steps at *TSR*=0.7

4.1.5 Computational validation

In order to ensure the CFD approach's accuracy in the present study, the similar model of [17] is made keeping all other design parameters same as [17] to determine the coefficient of performance (*Cp*) of the turbine model and computed at different *TSR* for a flow stream-speed (0.35) m/s. The comparison of the computational and published experimental results is shown in Fig.4.8. It is seen that CFD results are higher than the experimental results due to the present two-dimensional simulations as compared to experimental design, since 2D simulation overestimates the results of the experimental analysis. The similar results are also obtained in [15, 18]. Moreover, in CFD analysis, we can set the boundary conditions, but in experimental analysis, the boundary conditions and limitations may vary due to the uniformity in the flow, wall side effects, etc. This may create the differences in the results when compare to CFD study. Therefore, in the present study, the maximum error between the experimental and CFD data is 2.86%, which is below 5% [19], obtained at *TSR* 0.45. Hence, it is appropriate to estimate Savonius turbine performance using the suggested CFD approach.

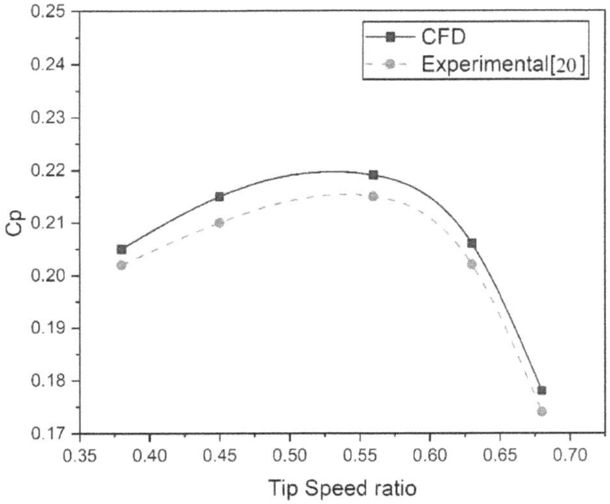

Fig.4.8 Validation of *Cp* of CFD study with [20] at different TSR

4.1.6 Taguchi Optimization

The Taguchi approach has been used to examine the combined impact of many parameters on the performance of the Model 3 turbine and, thus, to obtain an optimal combination of design variables. Model 3 turbine is considered for optimization as it has better performance than Model 1 and Model 2. For this investigation, the Taguchi analysis, an orthogonal approach, was used, and the procedure for performing this analysis can be found in the literature [20, 21]. The flow-speed, overlapping, side clearance, and winglet offset of the turbine are the design variables. Although flow-speed is considered as a design variable in the turbine development phase, after obtaining the optimized conditions, the impacts of other flow speeds will still be investigated to apply this design to any practical water reserve conditions. Table 4.3 shows information about the levels, controls, and factors. A, B, C, and D are the equivalent factors, and for each of these components, four distinct levels have been considered. It is necessary to carry out a total of 16 runs using the combination of the given experiments in an L16 (4×4) orthogonal array to achieve an optimal set of values.

Table 4.3 Details of levels, design parameters, and factors

Factor	Design parameter	Level			
		1	2	3	4
A	Flow-speed (m/s)	0.3	0.4	0.5	0.6
B	Overlapping percentage (%)	15	20	25	30
C	Side clearance percentage (%)	4.16	6.14	6.70	7.10
D	Winglet offset (mm)	10	15	20	25

Considering the different combinations of levels and variables, Table 4.4 displays the values of the S-N ratios and the mean of *Cp* in the L16 (4×4) orthogonal array. Further, Fig.4.9 (a) and (b) display the mean of the S-N ratio at various levels corresponding to the various factors, and it also displays the corresponding mean *Cp* values. It can be observed that level 2 shows the optimal design conditions (i.e. gap width ratio 6.14 %, overlapping 20 %, and offset 15 mm) along with 0.3 m/s flow-speed, as it has the lowest S-N ratio (i.e. −11.2009) and maximum mean *Cp* (i.e. 0.263).

Table 4.4 Details of the level combinations used in the L16 orthogonal array studies

Run	Factor				S-N ratio	Mean of *Cp*
	A	B	C	D		
1	1	1	1	1	-25.3521	0.054
2	1	2	2	2	-11.2009	0.263
3	1	3	3	3	-25.0362	0.056
4	1	4	4	4	-25.1927	0.055
5	2	1	2	3	-23.8764	0.064
6	2	2	1	4	-11.7568	0.258
7	2	3	4	1	-22.3837	0.076
8	2	4	3	2	-23.4785	0.067
9	3	1	3	4	-24.2934	0.061
10	3	2	4	3	-11.4373	0.235
11	3	3	1	2	-22.7335	0.073
12	3	4	2	1	-23.7417	0.065
13	4	1	4	2	-24.4370	0.06
14	4	2	3	1	-11.5024	0.225
15	4	3	2	4	-23.7417	0.065
16	4	4	1	3	-24.5830	0.059

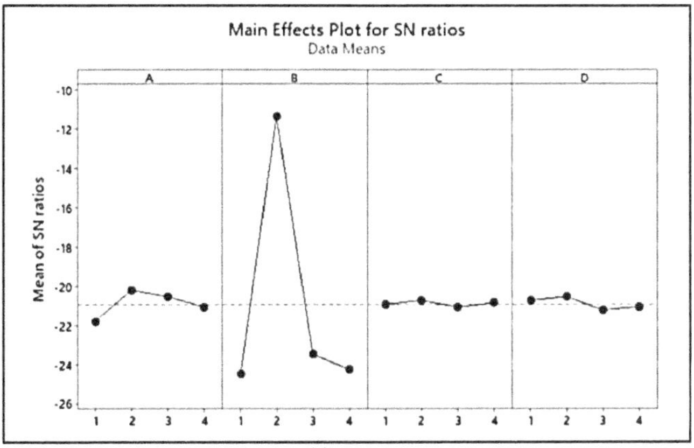

(a)

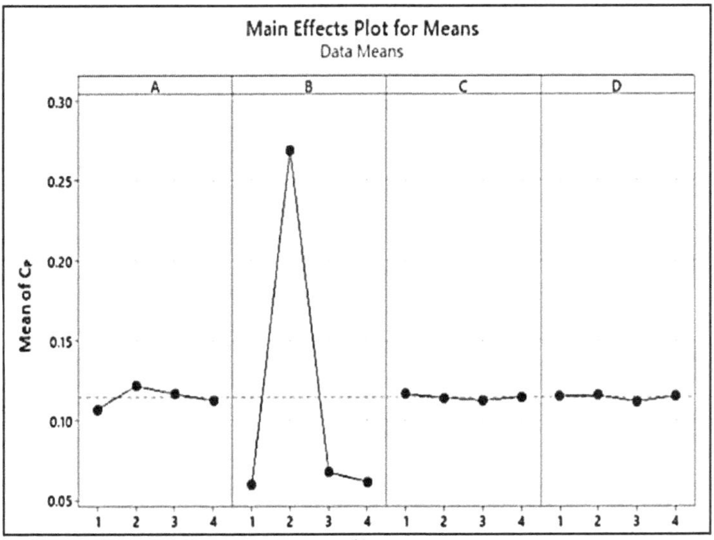

(b)

Fig.4.9 (a) Mean of the SN ratios for various levels and factors and (b) Mean *Cp* at various levels and factors

Reference

[1] NACA 63-415 AIRFOIL (n63415-il).(n.d.). http://airfoiltools.com/airfoil/details?airfoil=n63415-il

[2] Dincer, I. ed., 2018. *Comprehensive energy systems*. Elsevier.

[3] Nath, B., Biswas, A., Das, B. and Misra, R.D., 2025. Design and flow analysis of a Savonius-type hydro-kinetic turbine with hybrid-blade profile. *Environmental Progress & Sustainable Energy*, p.e70173.

[4] Roy, S. and Ducoin, A., 2016. Unsteady analysis on the instantaneous forces and moment arms acting on a novel Savonius-style wind turbine. Energy Conversion and Management, 121, pp.281-296.

[5] Talukdar, P.K., Sardar, A., Kulkarni, V. and Saha, U.K., 2018. Parametric analysis of model Savonius hydro-kinetic turbines through experimental and computational investigations. *Energy Conversion and Management*, *158*, pp.36-49.

[6] Osama, S., Emam, M., Ookawara, S. and Ahmed, M., 2024. Enhancing the performance of vertical axis hydrokinetic Savonius turbines using a novel cambered

hydrofoil profile for rotor blades. *Ocean Engineering*, *292*, p.116561.

[7] Riglin, J., 2016. *Design, modeling, and prototyping of a hydro-kinetic turbine unit for river application*. Lehigh University.

[8] Menter, F.R., 1994. Two-equation eddy-viscosity turbulence models for engineering applications. AIAA journal, 32(8), pp.1598-1605.

[9] Fluent, 12.0 Theory Guide, Ansys Inc, 2009.

[10] Meshing, S., 2011. Introduction to ANSYS Meshing. *Meshing Methods*.

[11] Basumatary, M., Biswas, A. and Misra, R.D., 2018. CFD analysis of an innovative combined lift and drag (CLD) based modified Savonius water turbine. Energy Conversion and Management, 174, pp.72-87.

[12] Celik, Y., Ingham, D., Ma, L. and Pourkashanian, M., 2023. Novel hybrid blade design and its impact on the overall and self-starting performance of a three-dimensional H-type Darrieus wind turbine. Journal of Fluids and Structures, 119, p.103876.

[13] Hashem, I. and Mohamed, M.H., 2018. Aerodynamic performance enhancements of H-rotor Darrieus wind turbine. Energy, 142, pp.531-545.

[14] Asr, M.T., Nezhad, E.Z., Mustapha, F. and Wiriadidjaja, S., 2016. Study on start-up characteristics of H-Darrieus vertical axis wind turbines comprising NACA 4-digit series blade airfoils. Energy, 112, pp.528-537.

[15] Dinh Le, A., Minh, B.D. and Trinh, C.D., 2022. High efficiency energy harvesting using a savonius turbine with multicurve and auxiliary blade. Journal of Fluids Engineering, 144(11), p.111207.

[16] Nakajima, M., Iio, S. and Ikeda, T., 2008. Performance of Savonius rotor for environmentally friendly hydraulic turbine. Journal of Fluid Science and Technology, 3(3), pp.420-429.

[17] Basumatary, M., Biswas, A. and Misra, R.D., 2021. Experimental verification of improved performance of Savonius-turbine with a combined lift and drag based blade profile for ultra-low head river application. Sustainable Energy Technologies and Assessments, 44, p.100999.

[18] Khan, Z.U., Ali, Z. and Uddin, E., 2022. Performance enhancement of vertical axis hydro- kinetic turbine using novel blade profile. Renewable energy, 188, pp.801-818.

[19] Tian, W., Mao, Z., Zhang, B. and Li, Y., 2018. Shape optimization of a Savonius wind rotor with different convex and concave sides. Renewable energy, 117, pp.287-299.

[20] Liao, C.N. and Kao, H.P., 2010. Supplier selection model using Taguchi loss function, analytical hierarchy process and multi-choice goal programming. Computers & Industrial Engineering, 58(4), pp.571-577.

[21] Taguchi G. Introduction to Quality Engineering. New York: McGraw-Hill; 1990.

CHAPTER 5

Computational and Experimental Investigations of the hybrid SAHT models

5.0 An Overview

This chapter presents the computational and experimental investigations of the proposed hybrid Savonius Hydrokinetic Turbine(SAHT). It discusses the computational investigations for different models (Model-1 to 3), CFD analysis of optimal model (Model-4), CFD analysis using different arc radii (Type-1 to 3) turbines under Model-4, and analysis of the experimental results for Model-4 under different design parameters at low flow speeds and low Tip Speed Ratio (TSR), which is commonly observed in perennial river stream current. The experiments were carried out in the recirculating open water channel for different low free-stream flow speed (≤1m/s). The detailed performance is evaluated in terms of torque (*T*), turbine power (P_{out}), coefficient of performance (*Cp*), torque coefficient(T_{coeff}) and turbine tip-speed ratio (*TSR*) for the SAHT models. The analysis of flow structure of the hybrid SAHT models are also discussed and comparison of CFD results with the experimental results is done in terms of *Cp*. Finally, the performances of the hybrid Savonius hydrokinetic turbines are compared with the previous published works, and final summary of the work is drawn at the end.

5.1 Computational study of the hybrid SAHT with combination of different parameters such as overlap ratio, gap width ratio, offset and cut out arc sections.

This section discusses the computational analysis of Model-1 to Model-4 with the different design parameters such as overlap ratio, gap width ratio, offset and cut out arc sections.

5.1.1 Primary computational findings on the variation of TSR on turbine performance

Fig. 5.1 represents the primary computational findings of torque (*T*), power (*P*), and *Cp* as a function of *TSR* (0.3–1) with 0.3 m/s flow-speed with only gap width ratio, no overlapping ratio, and offset (*Model 1*). It has been observed that torque, power, and *Cp* gradually rise to a maximum value of 0.069 Nm, 0.098 W, and 0.093, respectively, at *TSR* 0.6 and then fall as the *TSR* rises further. Fig. 5.2(a-d) depict the changes in torque, power, and *Cp* as a function of *TSR* (0.3–1) with 0.3 m/s flow-speed at overlapping ratio of 15 %, 20 %, 25 %, and 30 %, and gap width ratio of 4.16 %, 6.14 %, 6.70 % and 7.10 % (*Model 2*). The torque, power, and *Cp* gradually increase and reach their maximum value at 0.8 *TSR* and then fall gradually as the *TSR* rises further. Fig. 5.3(a-d) represent the torque, power, and *Cp* with combined overlapping ratio, gap width ratio, and offset of 10.0 mm, 15.0 mm, 20.0 mm, and 25.0 mm at 0.3 m/s flow-speed (*Model 3*). The torque, power, and *Cp* gradually increase and reach their maximum value at 0.7 *TSR* and then fall gradually as the *TSR* rises further. It can be observed that the maximum *Cp* obtained is 0.183 at 20 % overlapping ratio, 6.14 % ratio, and offset of 15 mm (Fig. 5.3b). The maximum torque and power obtained are 0.14 Nm and 0.22 W, and they also decrease as the gap width and overlapping ratio increase. Thus, it can be observed from Fig. 5.1, Fig. 5.2, Fig. 5.3 that the hydrodynamic performances with the addition of offset, overlapping, and gap width ratio are higher than the combination of overlapping and gap width ratio or without overlapping ratio, gap width ratio, and offset. Furthermore, as the overlapping and gap width ratio increase beyond 20 % and 6.14 %, respectively, the *Cp* value reduces. This is due to the bypassing of flow through these increased levels, resulting in lower thrust on the blades [1], [2]. However, from Fig. 5.3 (a-d), it is seen that as the *TSR* rises to 0.7, the turbine can still generate work, but as the *TSR* increases beyond 0.7 it gradually decreases regardless the flow speed and the related design conditions. This is because TSR higher than 0.7 becomes incompatible with the size of the turbine.

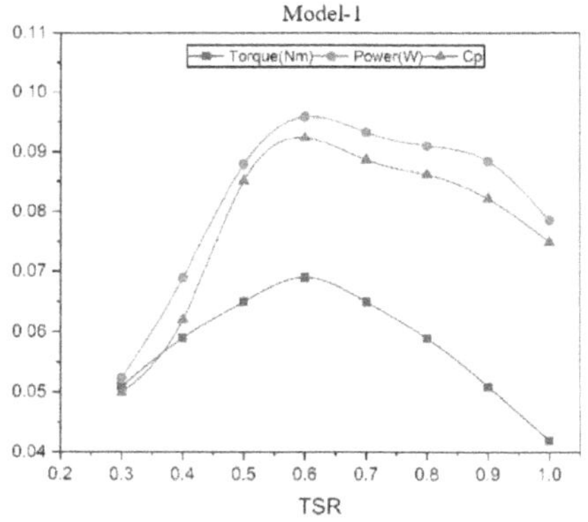

Fig. 5.1. Torque, power, and *Cp* variations with *TSR* at 0.3 m/s flow-speed without

overlapping and gap width ratio, and offset.

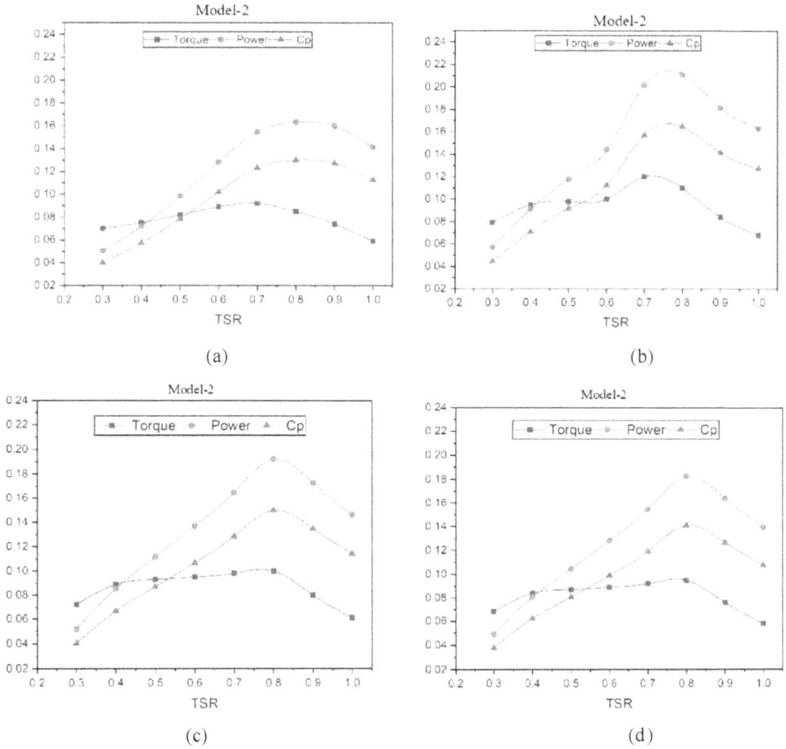

Fig. 5.2 Torque, power, and *Cp* variations with *TSR* at 0.3 m/s flow-speed with gap width and overlapping ratio (a) 4.16 %, 15 % (b) 6.14 %, 20 % (c) 6.70 %, 25 % (d)7.10 %, 30 %.

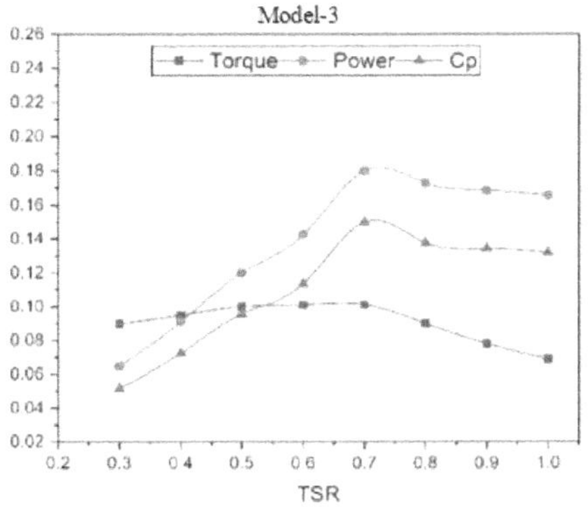

(a)

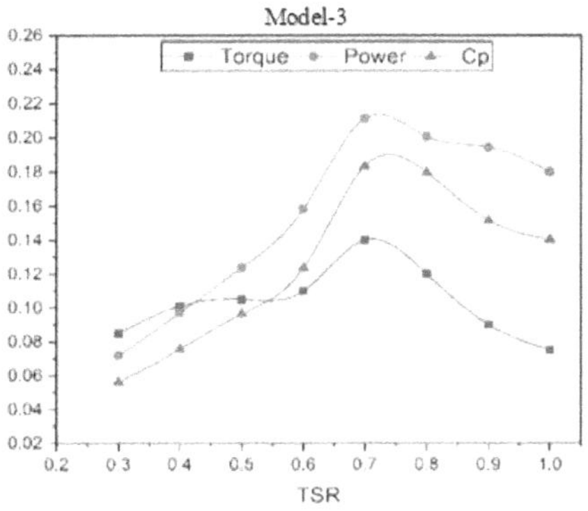

(b)

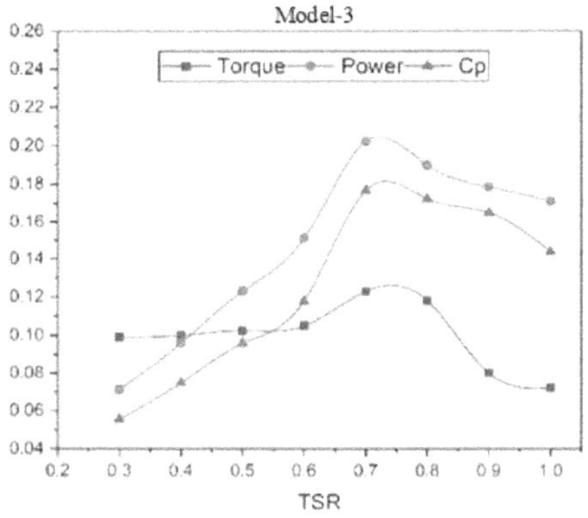

(c)

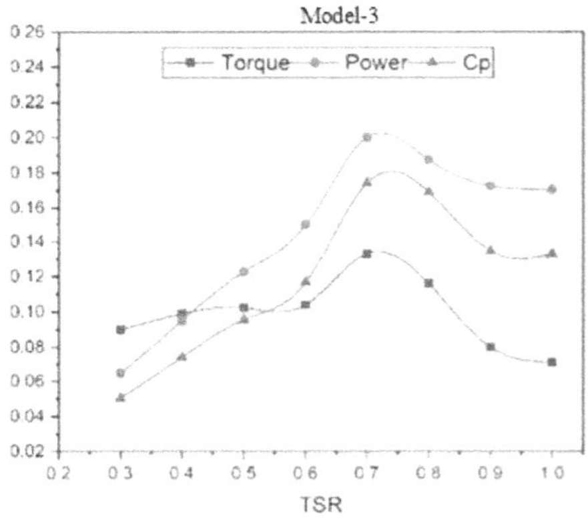

(d)

Fig.5.3 Torque, power, and *Cp* variations with TSR at flow-speed 0.3 m/s with gap width ratio, overlapping ratio, and offset of- (a) 4.16%, 15%, and 10 mm (b) 6.14%, 20%, and 15 mm (c) 6.70%, 25% and 20 mm and (d) 7.10%, 30%, 25 mm

From the above analysis, it has been observed that Model-3 has shown better performances than Model-1 and Model-2. Therefore, the optimization is performed for Model-3, which is discussed in the next section.

5.2 CFD analysis of optimized design(Model-3) and compared with Model-1 and Model-2

This section shows the detailed optimization results of Model-3 at low flow speeds (0.3-0.6) m/s and discusses the comparison of hydrodynamic parameters with Model-1 and Model-2 turbine designs.

5.2.1 CFD simulation details of optimized Model-3

The simulation outcomes, which evaluate the influence of several optimized design factors on the turbine, are listed in Table 5.1. It shows the results for each simulation values of the applicable operating conditions, such as flow-speeds and angle of attack (AoA) and employed design parameters i.e. offset, gap width, and overlapping ratio and the corresponding hydrodynamic parameters (such as lift coefficient (C_L), drag coefficient (C_D), torque coefficient (T_{coeff}) and Cp, which are obtained at 0.7 *TSR*, as maximum Cp is obtained at 0.7 *TSR* for the Model 3 (section 5.1.1). Table 5.1 depicts that with optimal design parameters and flow-speed 0.4 m/s for simulation run 2, higher hydrodynamic performances are attained by Model 3.

Table 5.1 Simulation details of hydrodynamic parameters

Simulation run	Design parameters			Operating conditions			Hydrodynamic parameters			
	Gap width ratio (%)	Overlapping ratio (%)	Offset (mm)	Flow-speed (m/s)	AoA (upstream-side blade)	AoA (downstream-side blade)	C_L	C_D	T_{coeff}	Cp
1	6.14	20	15	0.3	30	-30	0.39	1.820	0.162	0.105
					45	-45	0.45	1.856	0.283	0.214
					60	-60	0.53	1.915	0.286	0.282
					90	-90	0.21	1.948	0.270	0.266
					120	-120	0.15	1.982	0.292	0.288
					150	-150	0.05	2.241	0.120	0.118
2	6.14	20	15	0.4	30	-30	0.131	0.369	0.112	0.110
					45	-45	0.152	1.086	0.222	0.218
					60	-60	0.167	0.819	0.335	0.330
					90	-90	-0.146	0.493	0.319	0.314
					120	-120	-0.247	0.015	0.305	0.301
					150	-150	0.154	0.005	0.177	0.174
3	6.14	20	15	0.5	30	-30	0.161	0.437	0.038	0.038
					45	-45	0.168	0.449	0.143	0.141
					60	-60	0.179	0.261	0.212	0.209
					90	-90	0.73	0.238	0.209	0.206
					120	-120	0.241	0.057	0.205	0.202
					150	-150	0.484	0.044	0.198	0.195
4	6.14	20	15	0.6	30	-30	0.187	0.435	0.029	0.029
					45	-45	0.135	0.586	0.059	0.058
					60	-60	-0.122	0.62	0.145	0.143
					90	-90	0.105	0.631	0.143	0.141
					120	-120	0.015	0.178	0.127	0.125
					150	-150	0.397	0.22	0.090	0.089

5.2.2 Study of flow-speed on turbine performance at optimal winglet offset, gap width, and overlapping ratio

Fig. 5.4(a-c) show the results of the hydrodynamic parameters such as *P*, *T*, and *Cp* of optimized Model-3 at concerning four flow speeds (0.3, 0.4, 0.5, and 0.6 m/s). For this optimal design combination, the hybrid SAHT performs great under flow-speed 0.4 m/s as per observations. The objective is to determine the impact of flow-speed over turbine torque at 0.7 *TSR*. This *TSR* value is considered in this section as the turbine power and *Cp* have been found to be better for this TSR in the previous section. Fig. 5.4(a) shows that the output power of the turbine increases when flow-speed reaches 0.4 m/s at 6.14 % gap width ratio. The hybrid SAHT operates with a maximum power output of 1.201 W at 0.4 m/s when gap width ratio between the blades is 6.14 %. Turbine power reduces when the flow-speed increases beyond 0.4 m/s. Fig. 5.4 (b) demonstrates that torque increases as the flow-speed increases to 0.4 m/s, then decreases as flow-speed increases further, which has a negative effect in terms of performance. The possible reason could be the design incompatibility for increased flow-speeds [3]. According to Fig. 5.4(c), *Cp* rises up to a maximum of 0.33 for a flow-speed 0.4 m/s; however, for increase in flow-speed beyond 0.4 m/s it declines monotonically in line with the same trend of torque variation. The benefits of keeping a 15 mm offset, 20 % overlapping, and 6.14 % gap width ratio are therefore clearly marked when the data are carefully examined from the obtained results. Table 5.2 provides a summary of *T*, *P*, and *Cp* for 15 mm offset, 20 % overlapping, and 6.14 % gap width ratio at 0.7 TSR with the investigated flow-speeds (0.3, 0.4, 0.5, and 0.6 m/s) to emphasize this point. All of the winglet-bladed hybrid SAHT's quantifiable performance metrics are depicted in Table 5.2.

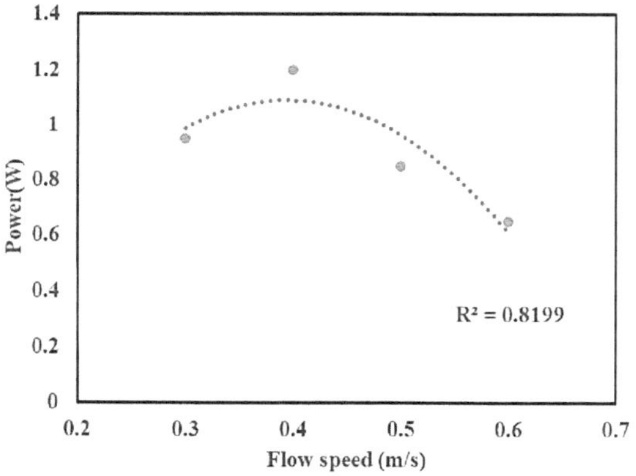

(a)

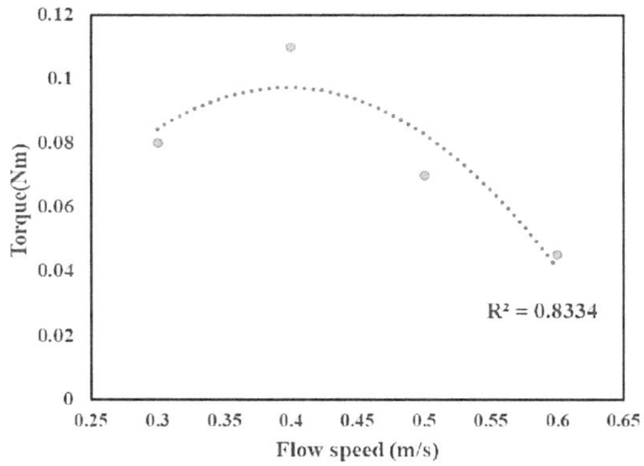

(b)

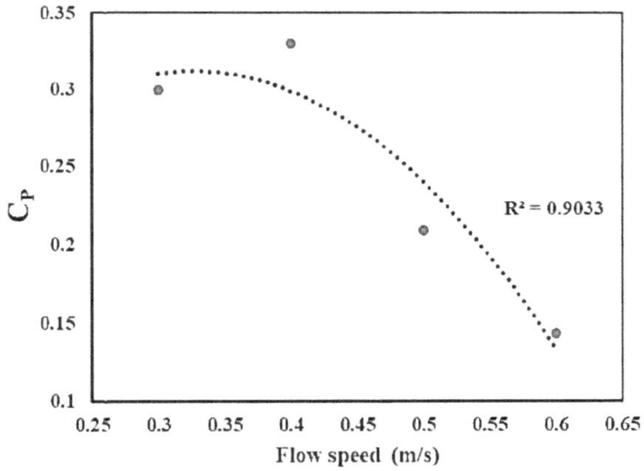

(c)

Fig. 5.4 (a) Power, (b) Torque, and (c) *Cp* at 0.7 *TSR* against flow-speeds of 0.4 m/s, 6.14% gap width ratio, 20% overlapping ratio, and 15 mm offset

Table 5.2. Computational results (at 15 mm offset, 20% overlapping, and 6.14% gap width ratio)

Flow-speed (m/s)	*T* (Nm)	*P* (W)	*Cp*
0.3	0.05	0.433	0.282
0.4	0.115	1.201	0.330
0.5	0.07	1.187	0.209
0.6	0.045	1.170	0.143

5.2.3 Variation of flow-speed on torque coefficient (T_{coeff})

Fig. 5.5 depicts the variation of T_{coeff} for the hybrid SAHT with flow-speed as 0.3–0.6 m/s at 0.7 *TSR*. Under various working situations, it displays the highest T_{coeff} that may be achieved, which represents the hybrid SAHT's ultimate torque performance attribute. The maximum T_{coeff} (0.335) is attained for the design condition of 6.14 % gap width ratio and 0.4 m/s flow-speed at a TSR of 0.7. The current turbine blades' 20 % overlapping ratio and gap width ratio of 6.14 %, and offset of 15 mm can accept maximum flow rates of up to 0.4 m/s with an optimal *TSR* of 0.7, allowing water to exert the possible force on the blade. However, as the flow-speed increases even further, the T_{coeff} decreases in line with the same trend of torque variation (Fig. 5.4(b)).

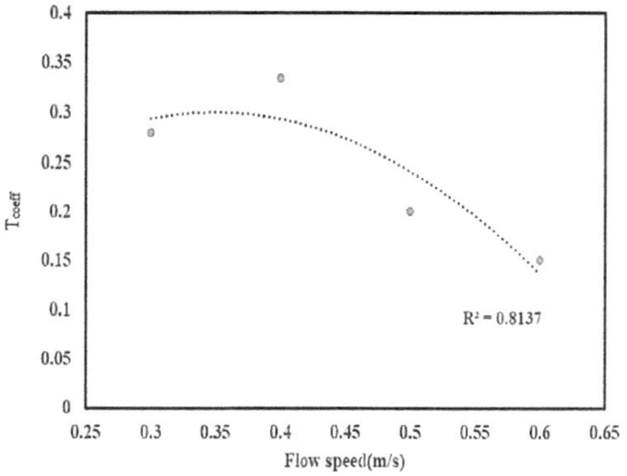

Fig.5.5 Variations of T_{coeff} for flow-speeds 0.3 m/s, 0.4 m/s, 0.5 m/s and 0.6 m/s

5.2.4 Turbulence intensity(T.I.) characteristics

Turbulence intensity, T.I (%), gives a good insight into the behaviour of the flow field across the turbine. The increase in T.I can also increase the lift and drag coefficient [4], [5], [6]. A comparison of the turbulence intensity is shown in Fig. 5.6(a-c) for Model 1, Model-2, and optimized model i.e. Model-3 at a flow-speed 0.4 m/s. In Model 1, the turbulence intensity through the turbine is maximum at the flow distance 50 mm, and then it drops to lower values within 50–225 mm, followed by another increase and decrease. This pattern of variation can be expected due to the unsteady aerodynamics of the turbine model. The maximum value of T.I is 13.28 % and the average turbulence intensity is found to be 3.25 %. In the case of Model 2, the amplitude of the turbulence intensity curve is higher, which is due to additional flow interactions with the turbine blades through gap width and overlapping. The maximum and average value of T.I, in this case, are 13.89 % and 3.51 %, respectively, which is quite higher than in Model 1.

But in Model-3, the turbulence intensity is higher than in Model 1 and Model 2. The maximum turbulence intensity reached 14.71 % with an average of 4.15 %. This is mostly due to the interaction of flow between the curved winglet and offset length of the straight blade section.

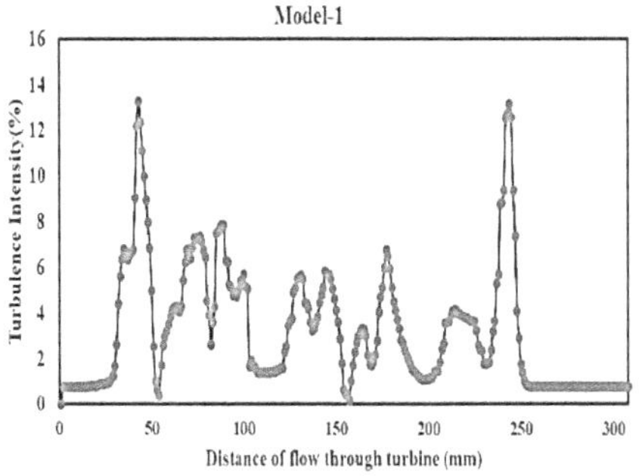

(a)

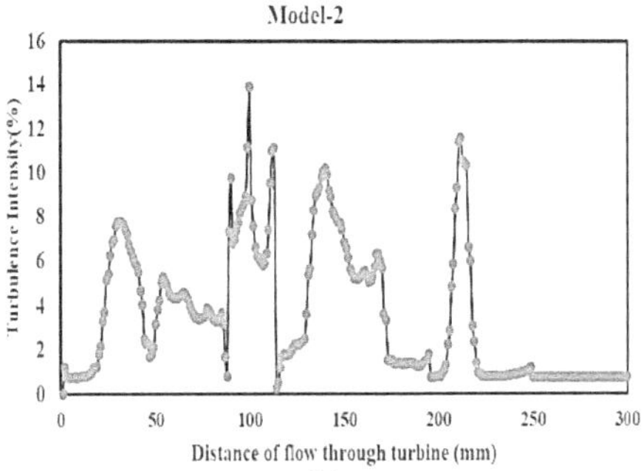

(b)

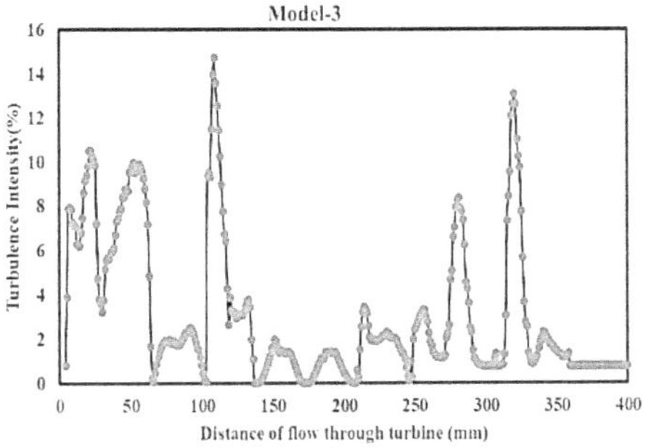

(c)

Fig.5.6 Turbulence Intensity for (a) Model 1, (b) Model 2, and (c) Model 3 with a flow-speed of 0.4 m/s.

5.2.5 Effect of turbulence intensity on turbine performances

The influence of turbulence intensity on turbine performances are shown in terms of lift coefficient (C_L), drag coefficient (C_D) and tangential thrust coefficient (C_{Th}) for Model 1, Model 2, and Model 3 at optimal flow speed of 0.4 m/s, as shown in Fig. 5.7(a-g). It is observed from the Fig. 5.7(a-c), that at lower turbulence intensities there is less effect on C_L for the three models. But as the turbulence intensity increases, the C_L also increases. Further, the same increases with the increase of attack angle up to a certain extent and then decreases with further increase of the latter. And Model 3 has shown maximum performances at attack angle (16°) with higher value of C_L, i.e. 0.88. The decrease in C_L for Model 3 is also minimum, which increases the stability of the curve winglet's dynamic characteristics [7]. It is also seen from Fig. 5.7(a-c); the stall angle also increases from Model 1 (12°) to Model 3 (i.e. 16°) signifying better performance of Model 3. The turbulence intensity has also influenced C_D for the three models, and it is higher in Model 3 as shown in Fig. 5.7(d). Higher values of these coefficients for Model 3 signify effective blade fluid interactions by its airfoil-shape based curved winglet and straight blade section. Similarly, the turbine performance in terms of C_{Th} is also estimated and depicted in Fig. 5.7(g) for Model 3. C_{Th} is directly linked with the torque and power performance of the turbine. It can be observed that as the turbulence intensity rises, C_{Th} also increases meaning turbine performance also increases. And interestingly, the negative fluctuations in C_{Th} are lowest at the highest turbulence intensity of 13 %, meaning loss of blade performance can be much averted with increased level of turbulence due to better blade-fluid interactions.

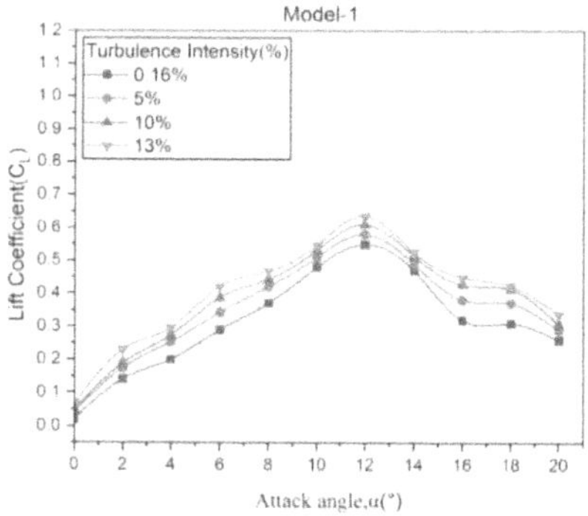

(a)

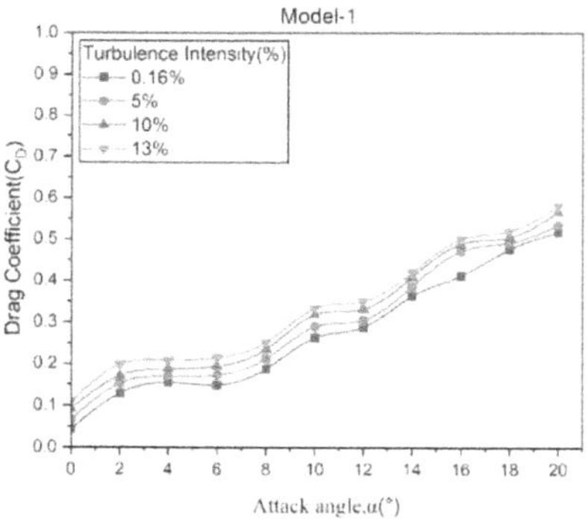

(b)

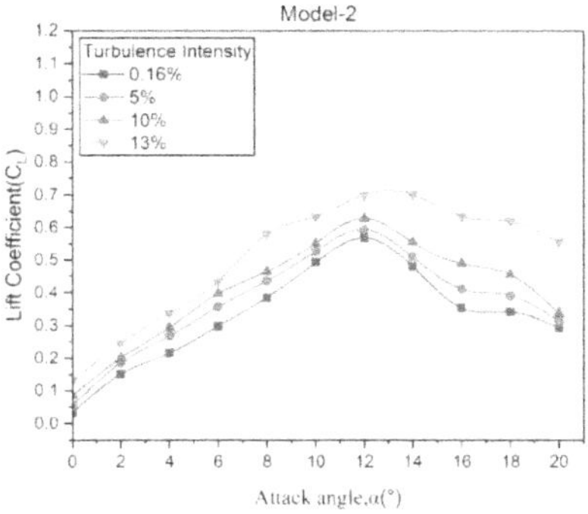

(c)

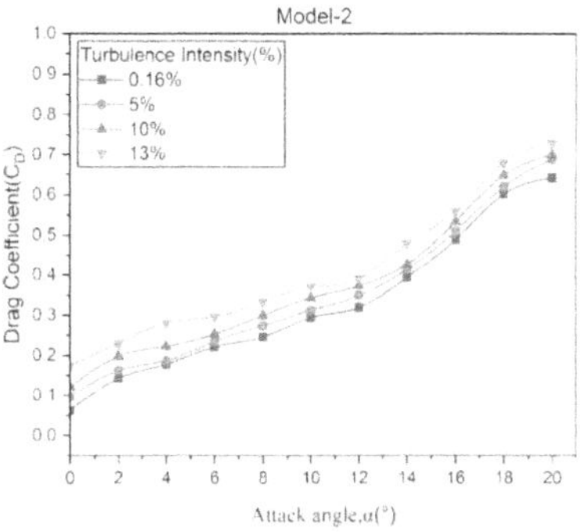

(d)

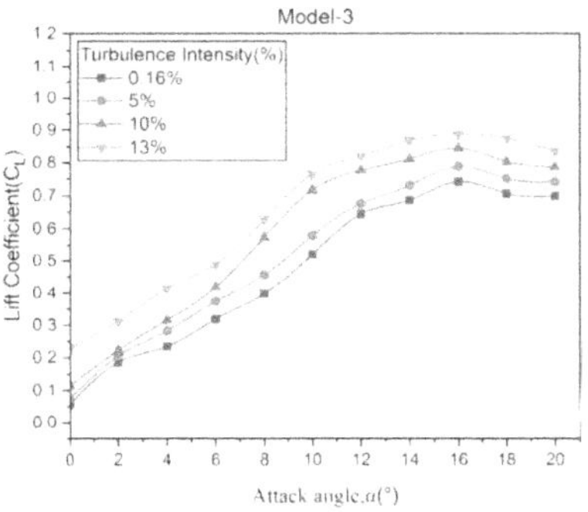

(c)

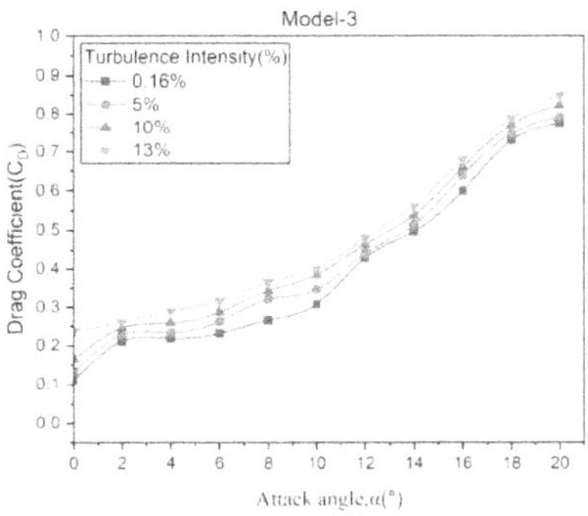

(f)

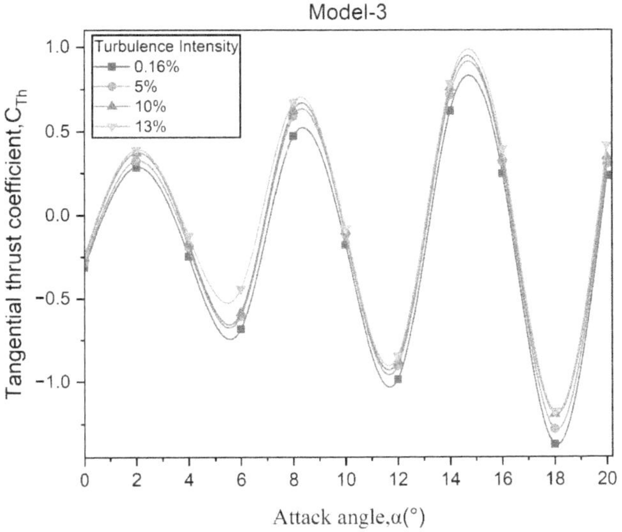

(g)

Fig.5.7(a-g) Turbulence intensity effect in terms of C_L, C_D and C_{Th} for Model 1, Model 2, and Model 3 with a flow-speed of 0.4 m/s

5.2.6 Pressure Characteristics

The static pressure around the three models for attack angles 135° and 360° are visualized in Fig. 5.8(a-b). In Fig. 5.8(a), for Model 3, at attack angle of 135°, high-pressure region sets up near the front-edge of the upstream-side blade and low-pressure region exists near convex face of the downstream-side blade, for which the static pressure difference is higher than Model 1 and Model 2. The Model 3 turbine generates more torque due to this higher pressure difference. Furthermore, near the overlapping zone, Model 3 undergoes through a high-pressure signature than Model 1 or Model 2, which is mostly attributed to the flow pressure from the offset section through the gap width. This physics will be deciphered from the flow physics analysis later on. At attack angle 135°, low-pressure zone grows near the tip of upstream-side blade at the trailing edge, but it is less for Model 3. At the attack angle of 360° as visualized in Fig. 5.8(b), across the overlapping portion, the resulting pressure is also increased in the case of Model 3. Moreover, the downstream-side blade has a better distribution for Model 3 and also uniform pressure distribution in the suction surfaces of both the blades. Overall, in all models, there is non-uniform pressure distribution across the pressure and suction surfaces causing loss of momentum (less in Model 3).

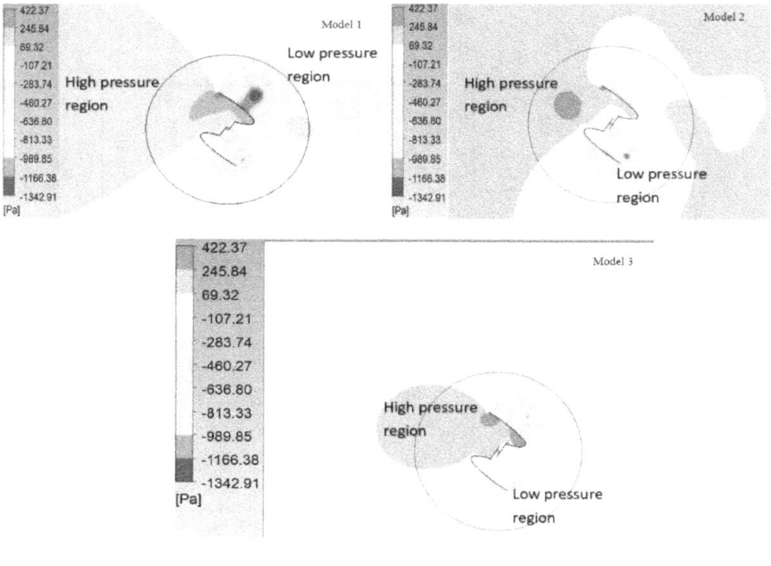

(a)

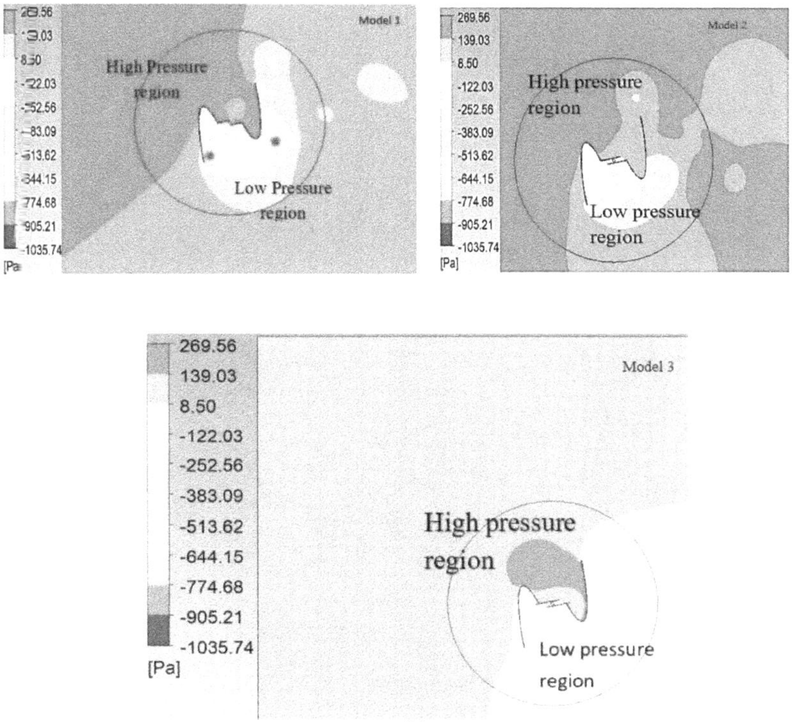

(b)

Fig 5.8. Static pressure signatures of (a) Models 1-3 at 135° attack angle and (b) Models 1-3 at 360° attack angle

5.2.7 Velocity Characteristics

An essential factor in determining the turbine's flow characteristics is the velocity signatures around the blades. Analysis of the velocity field comparisons among the three models is done. Fig. 5.9 (a and b) display the velocity signatures at the attack angles 135°and 360°. At 130° attack angle, as shown in Fig. 5.9 (a), a high-velocity zone is observed near the pressure surface of the upstream-side blade for the three models. Low-speed zones can be observed near the downstream-side blade in Model 1, followed by Model 2 and Model 3. But the loss of momentum (marked by blue patch zones) is lower in case of Model 3. Further, velocity increases at the tip of the blades at an attack angle of 360°, and it is greater in Model 3, as visualized in Fig. 5.9 (b), which will propel the turbine with a greater momentum.

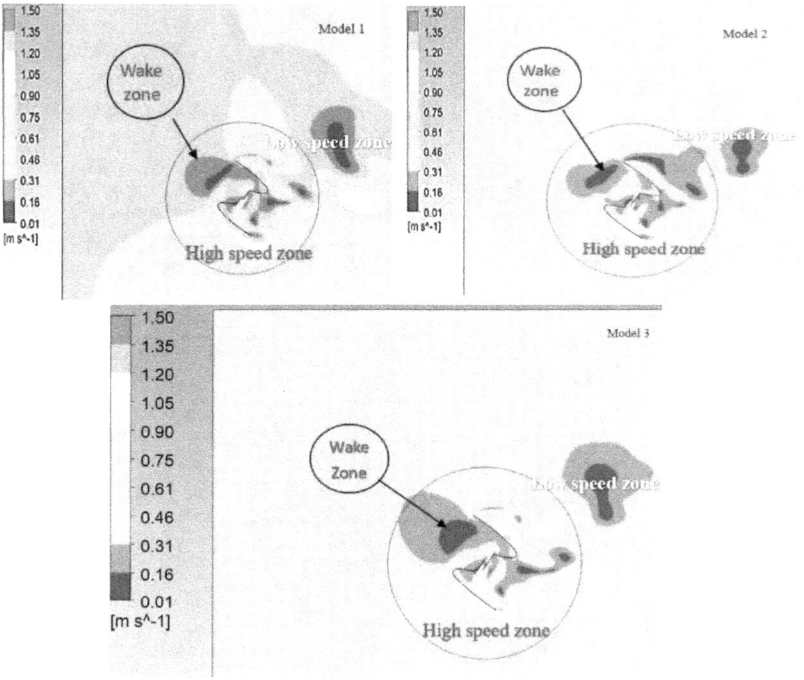

(a)

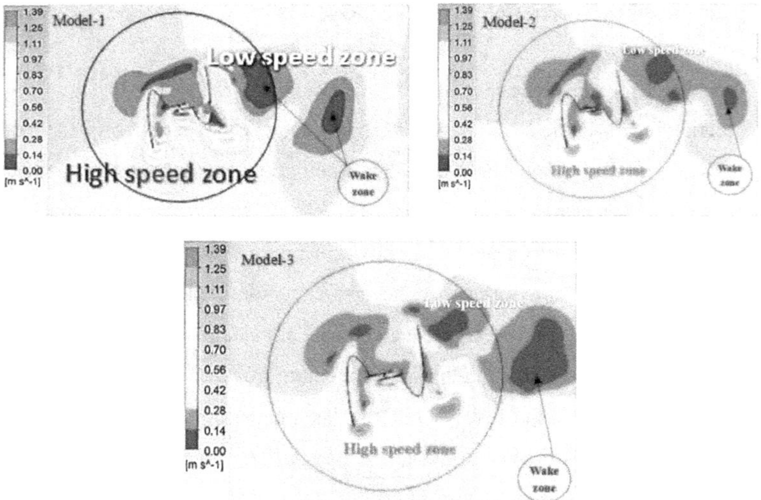

(b)

Fig.5.9 (a) Velocity signatures of- (a) Models 1-3 at 135° attack angle and (b) Models 1-3 at 360° attack angle

5.2.8 Vorticity characteristics

Vorticity, in simple terms, is a curl of the velocity vector. The vorticity signature at a *TSR* of 0.7 for an attack angle of 360° is shown in Fig. 5.10 (a-c). A vorticity ranges of 0 to 8420.08 (s−1) is observed. Vorticity reduces the performance of the turbine by obstructing flow from producing a thrust effect onto the blades. Low vorticity diminishes the obstruction for fluid to interact with blades, and as a result, the turbine performs better [8]. It is observed that high-intensity deep stall vortices zones near the pressure and suction side of the advancing and downstream-side blade for Model 1 and Model 2 are generated, which result in the formation of a low-pressure effect on the blades [9]. But in Model 3, the vorticity on the convex and concave faces of both the blades is comparatively less, which creates a high-pressure effect on the blades. The offset portion of the straight-blade sections in Model 3 rotor has minimum vortices than the other models. Moreover, the pressure sides of the curve winglet for the advancing and concave side of the downstream-side blade consist of low vorticity, which maintains the pressure on these sides to produce enough torque.

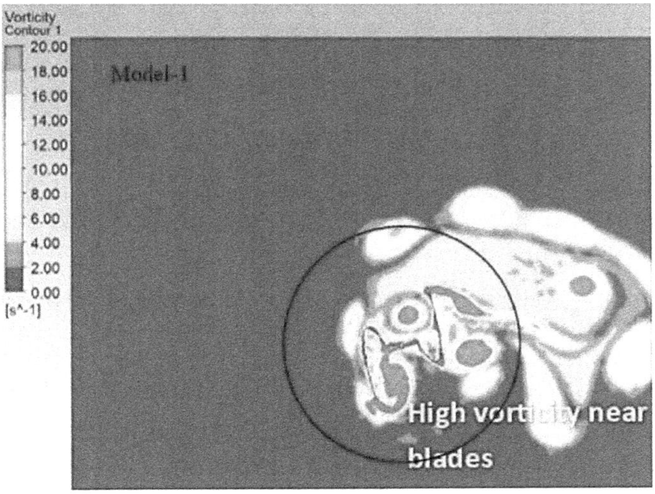

(a)

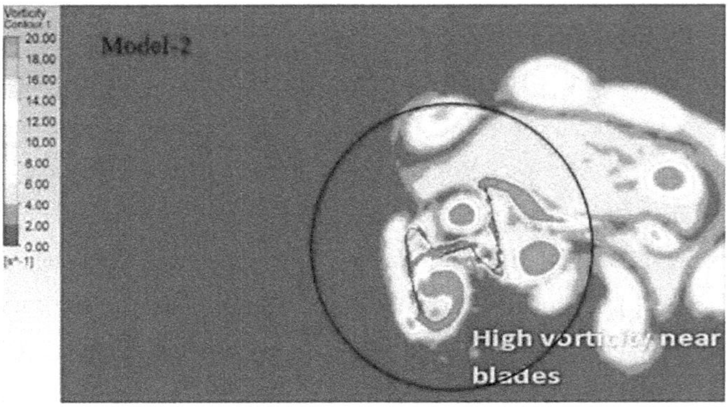

(b)

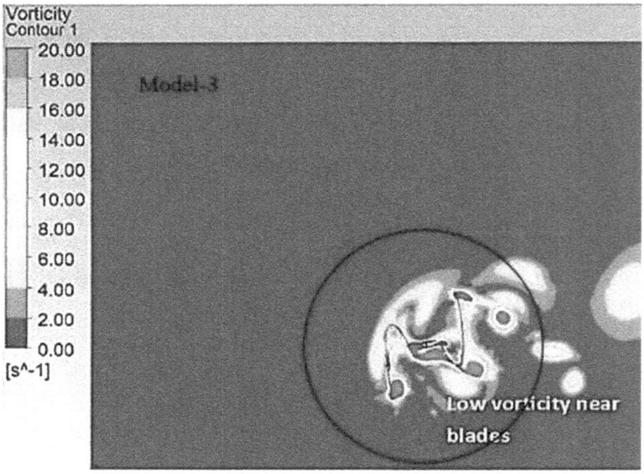

(c)

Fig.5.10 Vorticity signature of modified NACA 63415 (a) Model 1, (b) Model 2 and (c) Model 3 at 360° attack angle, respectively

5.2.9 Effect of hysteresis for lift and drag coefficient at different attack angles

The hysteresis effect is caused by variations in the hydrodynamic force with rising and falling attack angles [10]. Fig. 5.11(a) and (b) show that hysteresis occurs in Model 1, Model 2, and Model 3 with lift and drag coefficient variations at different attack angles. Hysteresis loops have emerged in a clockwise direction for the lift coefficient (C_L) for every model. The increasing pattern of C_L for all three models is similar at the beginning, but it decreases rapidly as the attack angle increases [4]. As stall happens at a higher attack angle (16°) in Model 3, C_L quickly touches a higher value of 0.88 than the other models. The design using offset with overlapping and gap width enables in minimizing the hysteresis loop's length, (16–20) ° in the lift coefficient of the Model 3 (Fig 5.11a). But the Model 1's hysteresis loop is extended over the angle of attack (12–20) °, which is due to early flow separation. As the angle of attack decreases, the maximum lift coefficient begins to decrease. The lift coefficients decrease by 13.6 %, 10.8 %, and 8.1 % for Model 1, Model 2 and Model 3, respectively. As a result, Model 3 can reduce the hydrodynamic hysteresis losses, increasing the stability of the curve winglet's dynamic characteristics [7]. Fig. 5.11(b) shows the hysteresis loop is emerged in anticlockwise [11]. And, the length of the C_D hysteresis loop is reduced in Model 3, as compared with the C_D hysteresis loop of the other models. Thus Model 3 is least affected by the hysteresis phenomenon.

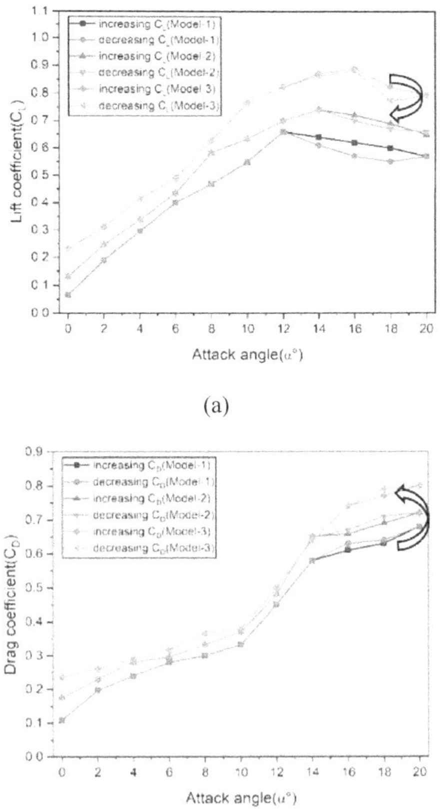

(a)

(b)

Fig.5.11 Variation of (a) C_L and (b) C_D for Model 1, 2 and 3 for flow-speed 0.4 m/s

The variation of lift coefficient, C_L for Model 3 with the flow-speed of 0.3, 0.4, 0.5, and 0.6 m/s, respectively, is shown in Fig. 5.12 (a). At a flow-speed of 0.3 m/s, the highest value of C_L is 0.81, whereas for flow-speed above 0.3 m/s, C_L is higher, and it is maximum (0.88) for flow-speed 0.4 m/s. The percentage increase in C_L is 8.64 %. However, the C_L with flow-speed of 0.5 and 0.6 m/s are lower than 0.4 m/s. Fig. 5.12(b) illustrates the C_D for flow-speed of 0.3, 0.4, 0.5, and 0.6 m/s, respectively. Here, C_D trend is similar as in Fig. 5.12(b). The C_D obtained from the graph (0.8) is higher than other flow-speeds.

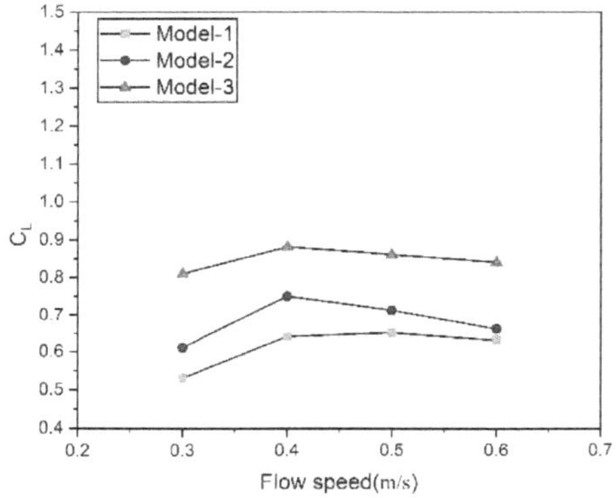

(a)

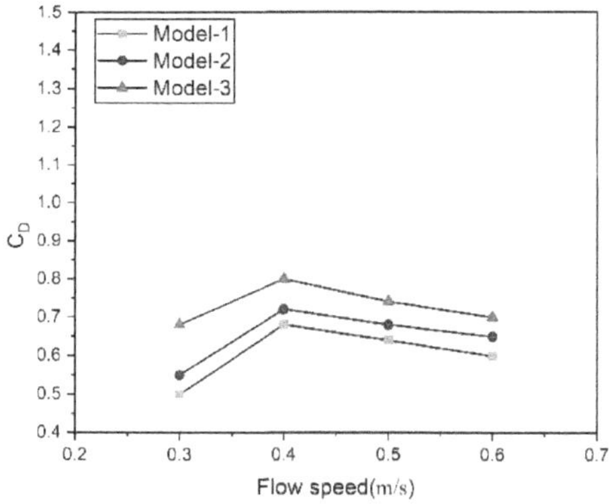

(b)

Fig.5.12 Variation of (a) C_L and (b) C_D for Model 1,2 and 3 for flow-speed (0.3-0.6) m/s

5.3 Effect of various cut-out arc sections of hybrid SAHT (Model-4) at optimal design conditions

Figure 5.13 visualizes the variations of hydrodynamic torque with flow time for different cut-out arc section percentages for optimized model, i.e. Model-3. The fluctuations of torque for γ = 11.5 %, γ = 12.5 % and γ = 13 % are generally higher up to flow time 6.5 s, but for γ = 12 %, there is a lower variation of torque and later at flow beyond 5.5 s. It can be seen that torque value is higher at γ = 12 % compared to γ = 12.5 % and γ = 13 %. This is due to the fluid interactions in the blade offset and overlapping zone, which reduces fluctuation in torque and negative torque as well. Therefore, γ = 12 % can be considered as the best cut-out arc section. With γ = 12 %, the new design Model-4 has been investigated, keeping the earlier obtained optimal design combinations of 6.14 % gap width and 20 % overlapping ratio, and 15 mm offset of Model-3.

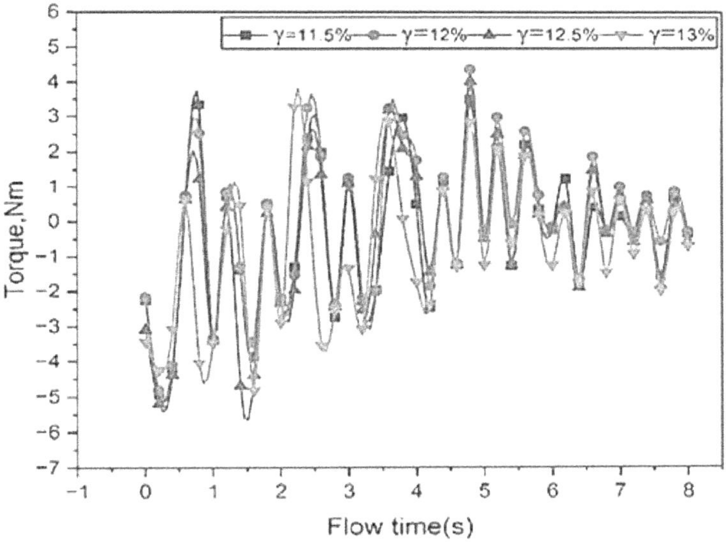

Fig. 5.13 Torque curve with respect to flow time(s) for four different cut out section at γ = 11.5 %, 12 %, 12.5 % and 13 %.

5.3.1 Hysteresis effect for Model 3 and Model 4 in terms of lift and drag coefficient

Fig. 5.14 (a) and (b) compare the curves of lift and drag coefficients of Model 3 and Model 4 for increasing and decreasing attack angles at a flow speed of 0.4 m/s. These curves show that hysteresis occurs in both Model 3 and Model 4, and loops have emerged in a clockwise direction for the lift coefficient for both models and onset of hysteresis is also the same i.e. at 16° attack angle (Fig. 5.14a). When stall happens at a higher attack angle of 18° than Model 3, C_L climbs to a higher value 0.99 (Model 4), Thus, stall is delayed and hysteresis loop size delayed for Model 4, resulting from the contribution of the improved design with the best cut-out arc section. Thus. Model 4 reduces the hysteresis losses, increasing the stability of the curve winglet's dynamic characteristics [7]. Fig. 5.14(b) shows the increment trend of the drag coefficient similar to Fig. 5.14 (b), and it shows the pattern of the hysteresis loop in anticlockwise.

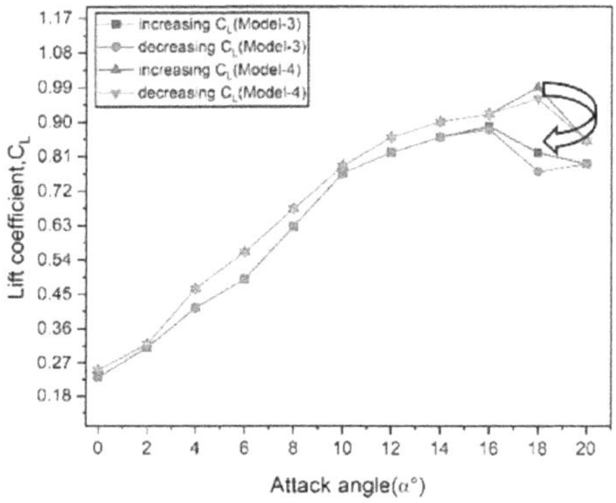

(a)

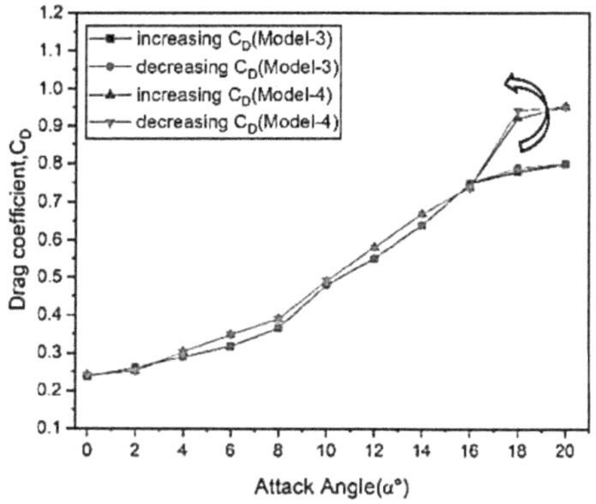

(b)

Fig.5.14 Variation of (a) lift coefficient and (b) drag coefficient for Model 3 and Model 4 at flow-speed 0.4m/s

5.3.2 Effect of *TSR* on coefficient of performance (*Cp*) and torque coefficient (T_{coeff})

As shown in Fig. 5.15 (a) and (b), the results of the measurement parameters (*Cp* and T_{coeff}) have now been obtained by plotting against *TSR* of 0.3, 0.5, 0.7 and 0.9 for Model 3 and Model 4 at 0.4 m/s flow-speed. Fig.5.15(a) shows that the turbine *Cp* increases as

TSR increases to 0.7, and after that, it decreases as *TSR* increases further. The hybrid SAHT operates at its maximum turbine *Cp* of 0.36 (for Model 4). The percentage enhancement in *Cp* as compared to Model 3 is 9.09 %. Fig. 5.15(b) demonstrates that T_{coeff} also increases up to 0.75 (for Model 4 and it is higher than Model 3) as *TSR* increases to 0.7, and then it decreases as *TSR* increases further. Therefore, Model 4 performs better at all *TSR* conditions (best at 0.7 *TSR*) for a flow-speed of 0.4 m/s. This also corroborates with the variation patterns of the un-optimized hybrid SAHT model.

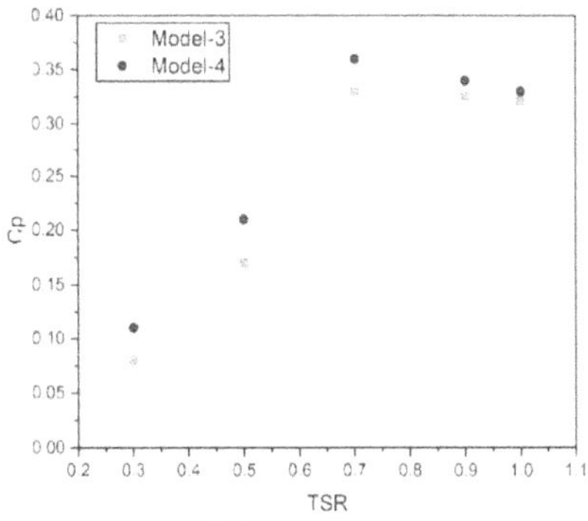

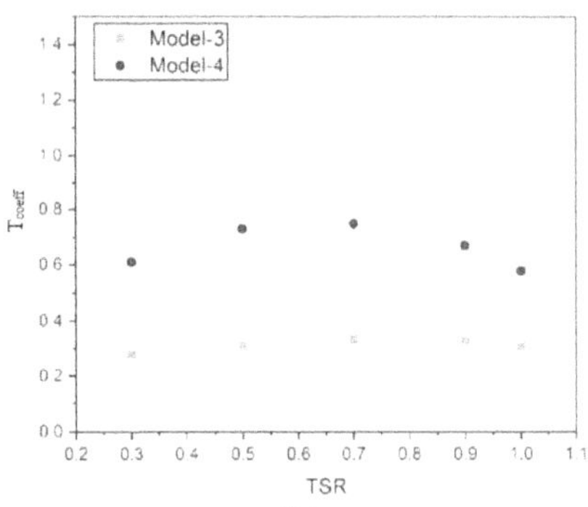

(a) (b)

Fig. 5.15 Signatures of (a) *Cp* and (b) C_T against the *TSR* (0.3, 0.5, 0.7, 0.9, and 1) for an optimal flow-speed of 0.4 m/s

5.3.3 Comparison of torque coefficient (T_{coeff}) for Model 3 and Model 4

Fig. 5.16 shows the comparison of T_{coeff} for Model 3 and Model 4 through one complete rotation with the help of a polar diagram at 0.4 m/s flow-speed. For the Model 4 turbine, higher values of T_{coeff} are achieved at angular positions $10° < \alpha < 80°$ and $155° < \alpha < 285°$ than the Model 3 angular positions $10° < \alpha < 85°$ and $155° < \alpha < 270°$. The torque coefficient of the Model 3 and Model 4, respectively, reaches peak values of 0.335 and 0.75 as the attack angle rises from 155° to 285° and 155° to 270° then begins to decrease. For the Model 4 turbine, a maximum value of T_{coeff} of 0.75 is obtained at an attack angle of 285°.

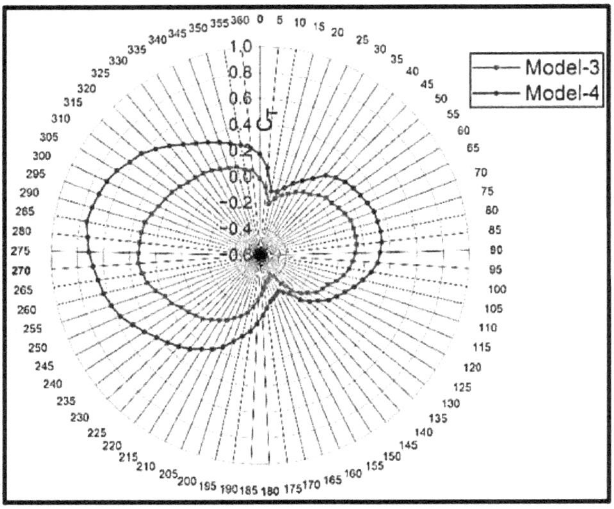

Fig. 5.16 T_{coeff} variation for both turbines at *TSR* 0.7 for one complete rotation

To underscore the insights of improved performance, the velocity vectors are captured across the blades of Model 4 in Fig. 5.17 at 135° attack angle for flow-speed 0.4 m/s and *TSR* 0.7. It can be observed that offset portion contributes at large followed by the gap width and overlapping. The offset portions improve the intensity of blade-fluid interactions, as there is a progression of Coanda-like vectors from the zone of offset-curve portion of the downstream-side blade towards the upstream-side blade through the gap width thereby pounding increased pressure on the upstream blade's straight section. Then, overlapping flow is also well developed within the overlapping zone. Further, offset-curve zone of the upstream-side blade improves the velocity signature on the suction surface of the curved winglet, thereby enhancing performance. Furthermore, it can be carefully observed that the flow is well-developed on the suction and pressure surface of the curved airfoil of upstream-side blade, whereas there is hydrodynamic loss (loss of momentum) from curved blades for Model 3 (see Fig. 5.9). Then much more uniform velocity distribution across the curved airfoil of both the blades in Model 4 can be seen in Fig. 5.17 than Model 3 (Fig. 5.9). Thus, it can be concluded that best cut-out arc section length of Model 4 results in improved performance of the hybrid SAHT than Model 3.

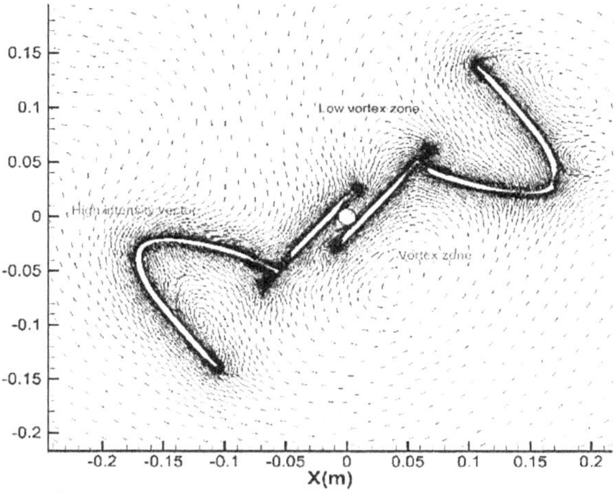

Fig.5.17 Velocity vector of Model 4 at attack angle 135°

5.3.4 Comparison of skin friction coefficient, C_f

For Model 3 and 4, the skin friction coefficients (C_f) are compared along the chord length of the curved winglet to obtain information regarding the separation point in Fig. 5.18 (a-c) at different attack angles. The wall-shear-stress reaches zero at the points where ($C_f = 0$), which means flow is on the verge of separation. The optimal cut-out arc section of the curve airfoil has a considerable effect on the flow in Model 4 as the flow remains attached with the blade for most of the chord length positions irrespective of attack angle than Model 3. Further, Model 3 has the tendency to reach to the separation state much earlier than Model 4.

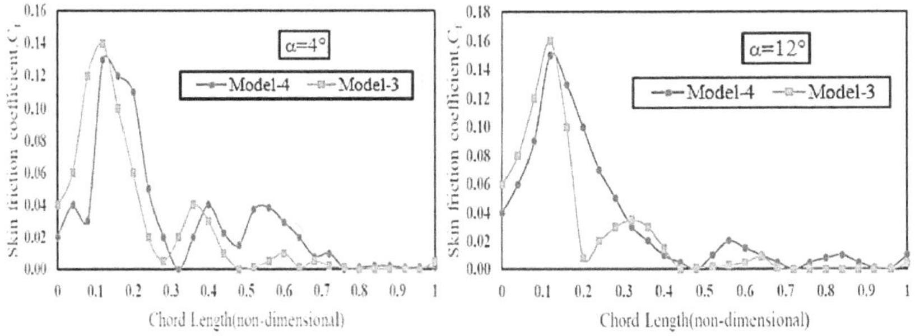

(a) (b)

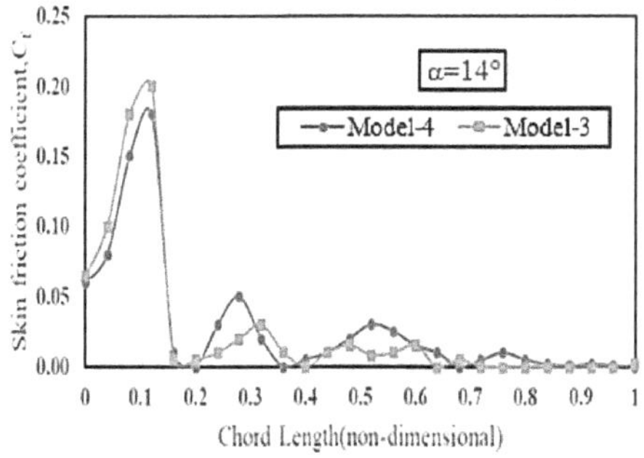

(c)

Fig. 5.18 C_f along curve winglet for flow-speed 0.4 m/s at (a) α = 4° (b) α = 12° (c) α = 14°

5.3.5 Q-criterion characteristics

A Q-criterion approach was used to depict vortex shedding [8], [12], [13], [14], [15], for the present hybrid SAHT Models. Q-criterion coloured by velocity magnitude at attack angle 90° is compared between Model 3 and Model 4 in Fig. 5.19 (a) and (b) at flow-speed 0.4 m/s and TSR 0.7. From Fig. 5.19(a) it can be seen, there is a long vortex ribbon separating from the suction surface of curved airfoil of upstream-side blade (Model 3), whereas flow holds onto the suction surface much better in case of Model 4 (Fig. 5.19b). This is attributed to the improved cut-out arc section and best airfoil arc angle of Model 4. Further, for the suction conditions in Model 3, vortices absorb energy from the mean flow much more than Model 4. As it can be seen in Fig.5.19(a), more separated vortices are prevalent across the straight section and the overlapping zone, whereas it is much less in case of Model 4 (Fig. 5.19b). Further, low velocity region is prevalent on the suction surface of the downstream curved airfoil blade for Model 3.

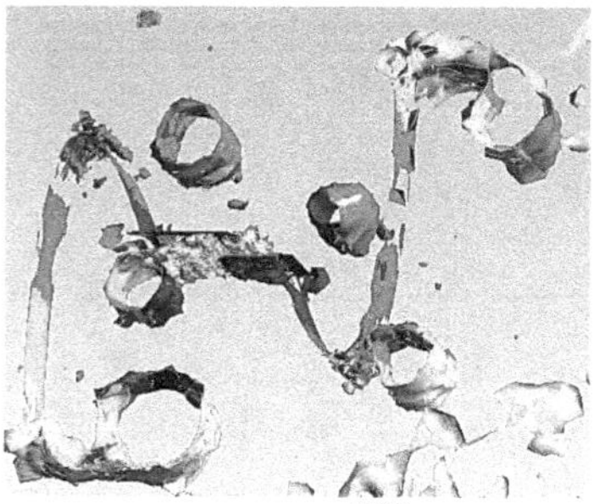

α=90°

(a)

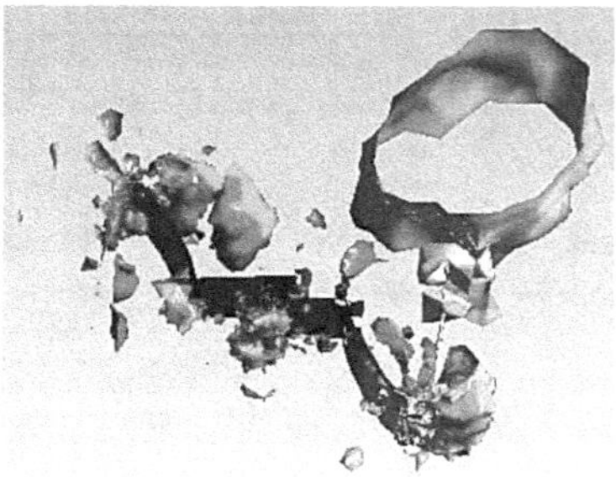

α=90°

(b)

Fig.5.19 Q-criterion coloured by velocity magnitude at various angle of attack for (a) Model 3 and (b)Model 4

5.3.6 Vorticity characteristics

The flow pattern of the experimental model(Model-5) of section 5.7 is shown in Fig. 5.20(a-d). It displays the vorticity signature of *SAHT* for ε = 6.14% and 7.10% with an attack angle of 45° and 110° at flow stream-speed of 0.8 m/s. There is an observed vorticity range of 0 to 20.0 (s^{-1}). If vorticity is less, then it improves turbine's performance by reducing the barrier for fluid to contact with the blades [31]. In Fig.5.20(a) and (b), i.e., at α=45°, the high stall vortex zone is observed adjacent to the tip of the advancing and returning blade and along the side of the flat straight blade in Fig.5.20(b) for ε =7.10%, but for ε =6.14% in Fig.5.20(a), at same attack angle low stall vortex zone is observed. The higher intensity stall vorticity patches adjacent to the pressure and suction sides of the upstream and downstream-side blades in ε =7.10%, causes low-pressure effects on the blades [9]. In Fig.5.20 (c) and (d), vorticity patches are reduced near blade side gap and overlapping portion for 110° attack angle, which shows that at higher attack angles vorticity zone decreases and it is low for ε =6.14% in Fig.5.20(c) when compared to ε =7.10% in Fig.5.20(d). Furthermore, vorticity patches leave the turbine blade towards downstream at 110° attack angle in Fig.5.20(c). Therefore, the fluid interactions with the blades is higher for ε =6.14%. This can be further compared from the flow structure study in section 5.3.7.

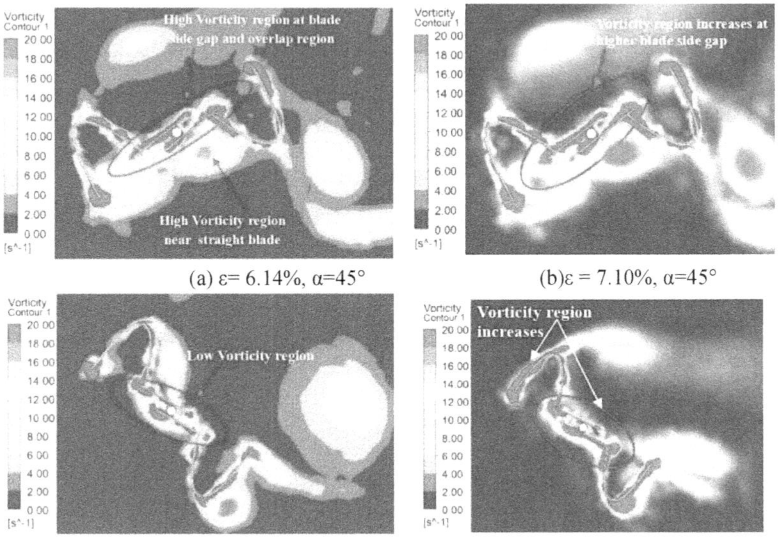

(a) ε= 6.14%, α=45° (b)ε = 7.10%, α=45°

(c) ε=6.14%, α=110° (d) ε=7.10%, α=110°

Fig.5.20 Flow structure signatures of *SAHT* at an attack angle (a) 45° and (b) 110° for ε=6.14% and 7.10%

5.3.7 Flow structure study

To study the flow interactions around the blades, the fluid flow characteristics at an attack angle 45° and 110° for the *SAHT* in terms of velocity vectors are shown in Fig.5.21 (a-d), respectively. In Fig.5.21 (a), the blade gap width and overlapping flow in *SAHT* supported by offset flow allows for additional flow thrust by guiding flow vectors to enter through the blade side gap and overlapped portions, imparting greater momentum towards the concave side of the advancing blade in the upstream direction. These underlying flow interaction features improve the torque, power, *Cp* and T_{coeff} of the *SAHT* blades for best design combinations of A_R=0.26, ε =6.14%, ψ=20% and 0.8 m/s stream-speed as obtained earlier. Supported by the offset, the flow enters through the gap width and overlap portion from the concave side of the returning blade and causes thrust to the advancing blade to rotate. But in Fig.5.21(b), as the gap width ratio increases (ε =7.10%), the flow vectors towards the concave side of the advancing blade decreases and passes through overlapped region without interacting properly with the blades, which causes recirculation zone near the straight-blades section. Thus, compared to flow characteristics at attack angle 110° in Fig.5.21(c) and (d), overlapping and blade gap width flow is similarly effective for *SAHT* as in case of 45° attack angle condition. The thrust on the blades' leading edges rises as a result of momentum restoration from the suction surfaces. The same is increased on rest of the blade surfaces. Thus, there are higher intensity velocity vectors at the leading ege of the advancing and returning blade for *SAHT*, and at the trailing edge of the returning blade,which implements more thrust from the mean flow for the blades of *SAHT*. But in Fig.5.21(d), the flow vectors through the blade gap width and overlapped portion again decreases at ε =7.10%. This implies that the SAHT can function better at ε =6.14% at α=45° and 110°.Therefore, in overall, due to all these effective flow controls for *SAHT* and maximum fluid interactions, *SAHT* with overlapping 20%, gap width 6.14% and offset 15 mm produces higher hydrodynamic performances for the present design of *SAHT*.

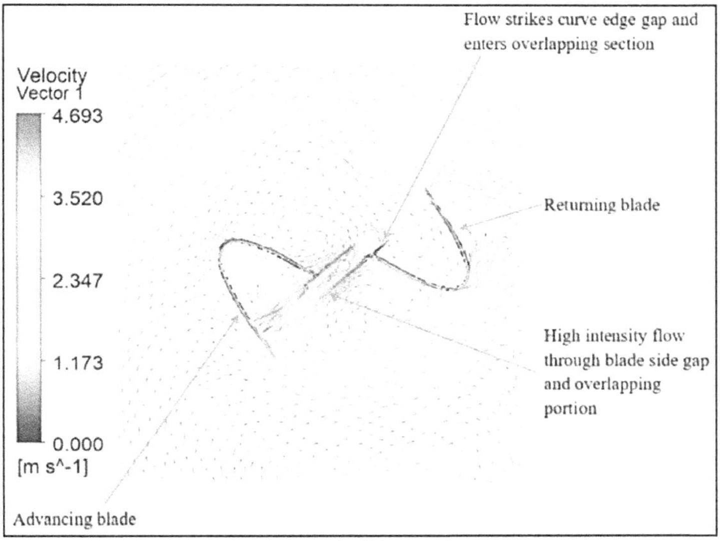

(a) $\varepsilon = 6.14\%$, $\alpha=45°$

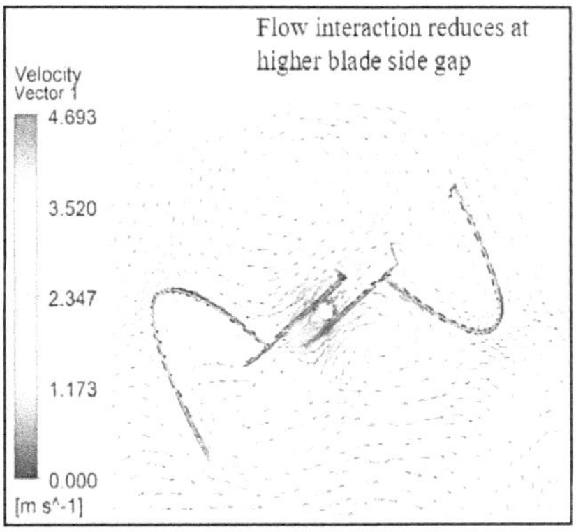

(b) $\varepsilon = 7.10\%$, $\alpha=45°$

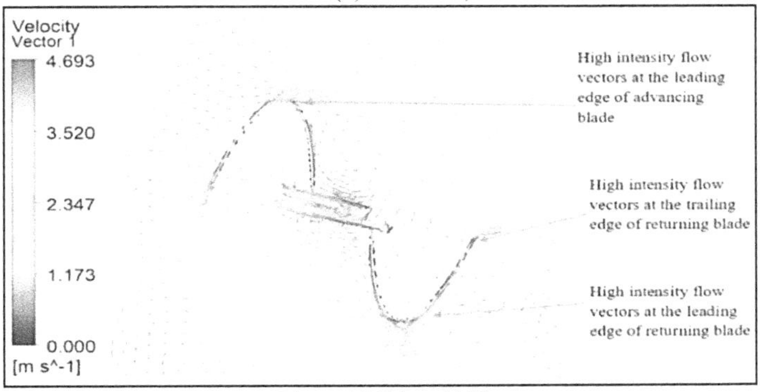

(c) $\varepsilon = 6.14\%$, $\alpha=110°$

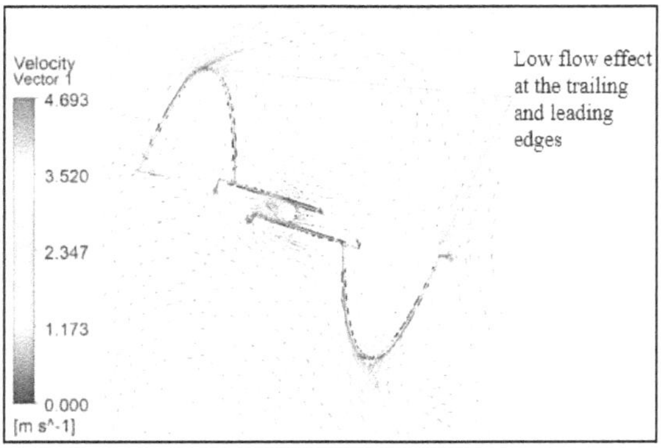

(d) ε = 7.10%, α=110°

Fig.5.21 Flow structure signatures of *SAHT* at an attack angle (a)45° and (b)110° for ε =6.14% and 7.10%

5.3.8 Comparison with the literature

The *Cp* of Model 4 are compared with the investigations of Khan et al. [9] and Basumatary et al. [16], as shown in Table 5.3, having almost similar flow-speed condition. The maximum *Cp* for Model 4 (0.36) is larger than the same (i.e., 0.15) obtained by [9], which is also more than [16]. The maximum percentage enhancement of *Cp* by Model 4 is 34.3 % when compared to [16].

Table 5.3 Comparison with the existing literature

Reference/Section no.	Year	Type of airfoil based design	Turbine Size D_T (mm)	Flow-speed(m/s)	*Cp*
Basumatary et al.[16]	2021	NACA 0018 airfoil-straight bladed	250	0.5	0.268
Khan et al.[9]	2022	S1048 airfoil-straight bladed	330	0.38	0.15
Model-4	-	NACA 63415 airfoil-straight bladed	334.62	0.4	0.36

5.3.9 Summary of the results

In the present work, two-bladed hybrid Savonius hydro-kinetic turbine (SAHT) models were designed, analyzed as well as optimized. In the first part of the study, performance of three different models were analyzed and Taguchi optimization with three design parameters such as overlapping (15–30) %, gap width ratio (4.16–7.10) % and offset

(10–25) mm was performed for Model 3 with operating flow-speeds (0.3–0.6) m/s to obtain the optimal design and operating conditions. In the second part, the cut-out arc sections and corresponding angles were varied in Model 3 and their best values were incorporated in hybrid SAHT (Model 4), and further comparisons were performed. The below mentioned are summarized as follows:

- The Model 3 SAHT performs better than Model 1 and Model 2 SAHTs, as it produces an output power 1.201 W, torque 0.115 N-m for operating flow-speed 0.4 m/s.
- The Model 3 SAHT generates a maximum *Cp* of 0.33 at an optimal overlapping and gap width of 20 % and 6.14 %, respectively, and an offset of 15 mm for the same flow-speed.
- The best SAHT model (i.e. Model 4) generates the highest *Cp* and T_{coeff} of 0.36 and 0.75, respectively at the same optimal design combinations of Model 3.
- Better hydrodynamic performance of Model 4 is contributed from the offset section of both upstream and downstream-side blades, as Coanda-like vectors from downstream progresses towards the upstream through gap width thereby imparting more pressure on the straight-blade section. Further, velocity vectors from offset section of the upstream blade improves the velocity signature on suction surface of the advancing blade.
- Model 4 SAHT has a better performance than Model 3 owing to well-developed flow patterns on the suction and pressure surface of the curved airfoil of upstream-side blade and more uniform velocity distribution across the curved airfoils of the turbine. Further, flow is less detached from the suction face of advancing blade and from the overlapping zone and straight section of Model 4.
- Airfoils parameters with best cut-out arc section length (12 %) and offset (15 mm) improve the SAHT performance more than the performance with traditional overlapping or gap width approaches (Model 2 and Model 3).
- When compared to the other curve profile of S, NACA, etc. based turbines previously studied, the performance coefficient of the present SAHT is improved.
- The SAHT with the best blade gap width ratio (6.14%) and overlapping ratio (20%) exhibits well-developed flow structures, leading to a greater flow thrust. This is enhanced by the offset and blade gap width flow and suitably covered by

the adjacent overlapping straight-blade sections. Additionally, *SAHT* turbine blades experience increased thrust due to high intensity shear layers at the leading and trailing edges of the advancing and returning blades.

- The vortical flow patterns around *SAHT* for gap width ratio 6.14% are less pronounced and do not completely adhere to the blades, which leads to improved stall regulation. The recirculation area on the suction side of the returning blade and the suction side of the advancing blade is diminished at 110° attack angle, allowing for the recovery of lost momentum and converting it into extra momentum for SAHT.

5.4 Computational study of the effects of different arc curvatures radii on hybrid SAHT with the optimal design conditions.

In this part, computational analysis for Model-4 have been performed by introducing the three different arc curvature radii to form three new hybrid turbine types, i.e., Type-1 (R_1=13 mm), Type-2 (R_2=15 mm) and Type-3 (R_3=17 mm). Initially, the results of the torque are analyzed for Type-1, 2 and 3 turbines using three aspect ratios (discussed in Chapter 3) for a low flow stream-speed of 0.5 m/s to obtain the best turbine design. Then, flow physics in terms of pressure, vorticity, velocity, etc., is compared to visualize the flow around the three turbine types.

5.4.1 Variation of torque (Nm) with flow time(s) for the three types of hybrid SAHTs

The torque curve profile for three turbine types at different A_R values, such as 0.48, 0.51, and 0.54 for low flow stream-speed (0.5 m/s), are given in Fig.5.22 (a-c). In these plots, flow time up to 2 secs during the start-up phase of rotation corresponds to 0.83 rpm. After 2 secs, the torque fluctuation becomes stable as the rotation attains quasi-static conditions state. It can be seen from Fig.5.22 (a) that the average fluctuation of torque at 0.51 A_R is lower than at 0.48 A_R and 0.54 A_R for Type-1 SAHT. This indicates that the 0.51 A_R design in the Type-1 SAHT turbine is better, and the peak of the torque curve is also greater than the other aspect ratios. In Fig.5.22 (b), the torque curve profiles of Type-2 SAHT with A_R value (0.54) show maximum torque fluctuation like Fig.5.22 (a), and the torque fluctuation for Type-2 SAHT is lower at 0.51 A_R. The fluctuation of torque is between -1.1 to 1.2 Nm for 0.51 A_R, and the amplitude of torque is also high throughout the flow time from 2-5 seconds. The average torque for Type-2 SAHT is higher than Type-1 and Type-3 at 0.51 A_R. Figure 5.22 (c) also shows the lower torque fluctuations at 0.51 A_R. However, the average torque in Type-3 SAHT is lower than Type-2 SAHT (in Fig.5.22 (b)), which indicates that the Type-2 SAHT design with 0.51 A_R is suitable for further analysis to determine the various hydrodynamic parameters.

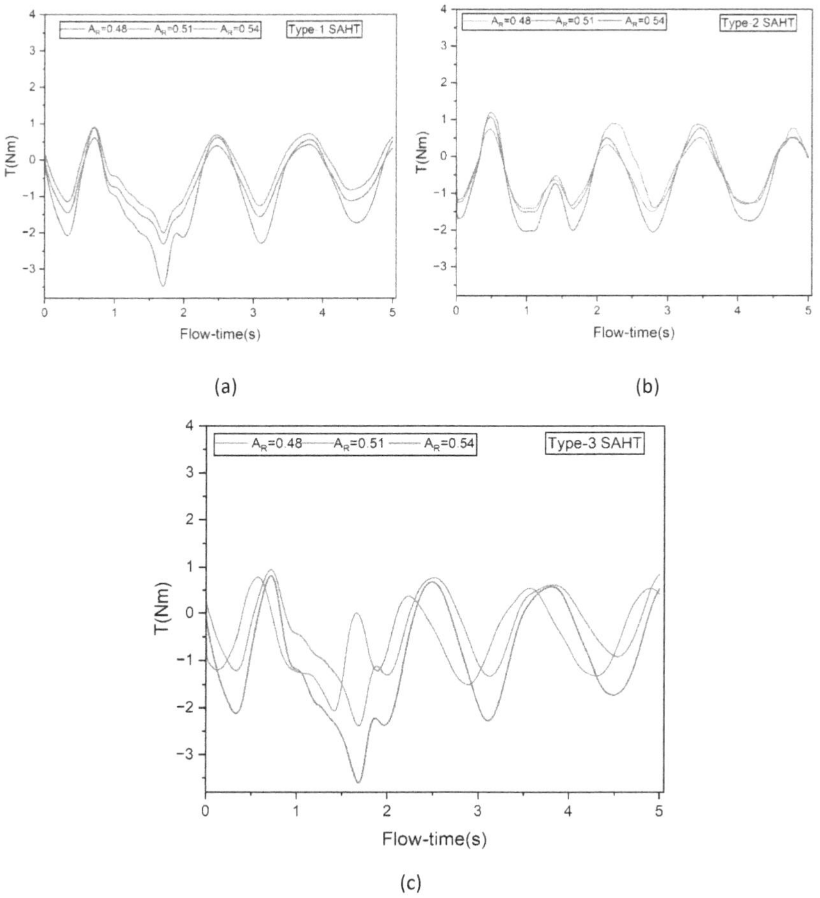

Fig.5.22 Comparison among (a) Type-1, (b) Type-2 and (c) Type-3 SAHT's with different aspect ratios at flow stream-speed 0.5 m/s

5.4.2 Pressure coefficient for the three types of SAHTs

To visualize the performances of the three types of SAHTs, the pressure coefficients (P_c) with respect to blade positions at an attack angle of 45° at A_R =0.51 are plotted in Fig.5.23(a-c), and the blade 1 and blade 2 indicate advancing and returning stages during rotation. From Fig.5.23(a) and 10(b), it is seen that the pressure coefficient distributions for blade 1 for Type-I SAHT are higher than Type-1 SAHT, and there is also an improvement of pressure coefficient distributions in the overlapping and blade gap width region. It is also observed that the pressure coefficient's loop area for blade 2 is greater for Type-2 SAHT, and more positive pressures are generated on the pressure side of the returning blade 2 for Type-2 than Type-1 and Type-3 SAHT. All these positive effects of pressure distributions can enhance the overall performance of Type-2 SAHT. The reason could be accounted for the effective flow control from the curved blade part towards the axis for the Type-2 SAHT, and due to overlapping and offset, the flow exerts more pressure. Overall higher average torque is generated in Type-2 SAHT (section 5.4.1). This is due to higher-pressure coefficient distributions across its blades. In Fig.5.23(c), the pressure coefficient's loop area decreases for blade 2 compared to Type-2, and the pressure coefficient difference is also reduced in the blade gap width and overlapping region, which is probably due to the reduction of arc radius of curvature. The reduced arc radius of curvature fails to accommodate the proper fluid interactions, resulting in a decrease in overall pressure coefficients during rotation. The flow physics in terms of static pressure contour, velocity contour, vorticity contour, and velocity vector are discussed in detail to express the turbine flow features further.

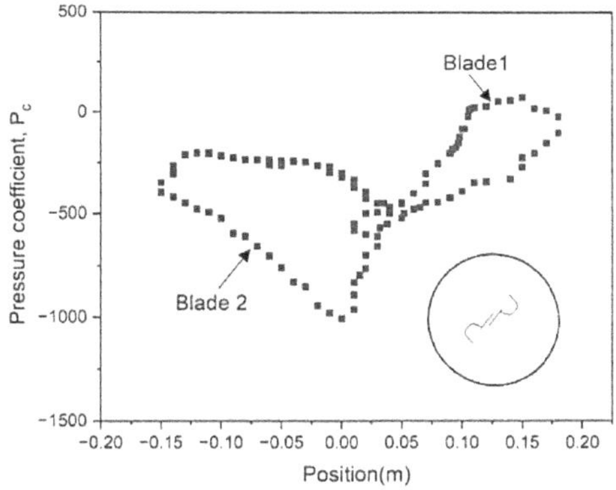

(a)Type-1 SAHT

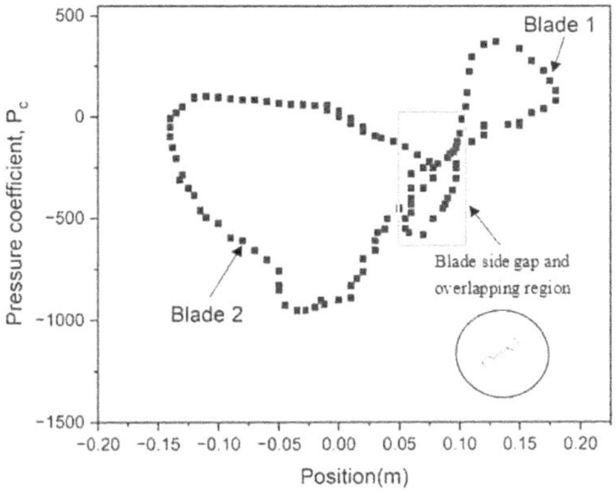

(b) Type-2 SAHT

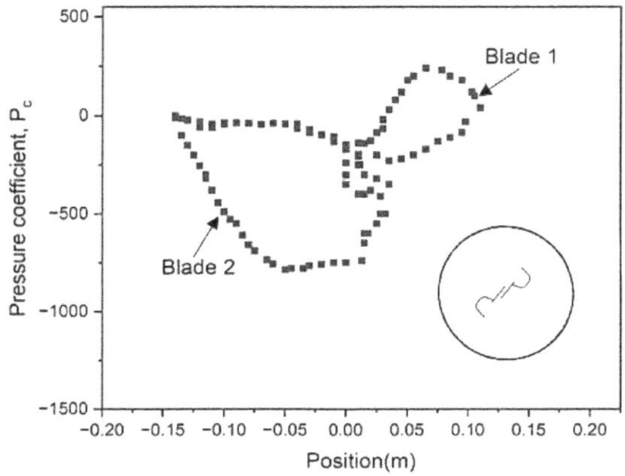

(c)Type-1 SAHT

Fig.5.23 Comparison of pressure coefficients for (a) Type-1, (b) Type-2 and (c) Type-3 SAHT's at 45° attack angle

5.4.3 Static pressure signatures

Figure 5.24(a-f) shows the static pressure surrounding the Type-1, 2 and 3 SAHTs at attack angles of 0° and 45° for 0.5 m/s flow stream-speed. In Fig.5.24 (a), i.e., for Type-1 SAHT, at 0° and 45° attack angles, a high-pressure region is seen at the concave side of the advancing blade, and a low-pressure region is observed near the convex side of the advancing blade. This pressure drop surrounding the rotor causes the turbine blades to start rotating, enabling the Savonius hydrokinetic turbine to generate electricity from the flowing water in the channel. A similar phenomenon was also observed in [17]. Also, the pressure intensity is more negatively distributed near the overlapping and side gap region of Type-1 SAHT at α=0°, and the low-pressure zone gradually increases for α=45°. For the Type-2 turbine in Fig.5.24(b), there is a larger pressure difference between the concave and convex sides of the advancing blade, resulting in the generation of more torque. Moreover, the resulting pressure is increased near the overlapping and side gap regions in SAHT. This follows similar insights into pressure distribution in the Type-2 turbine, as seen in section 5.4.2. At a higher attack angle of 45° for Type-2 SAHT, pressure on the suction surface of Blade 2 is increased. In Fig.5.24(c), at 0° attack angle, negative pressure is seen on the returning blade 2 and pressure difference across blade 1 is lower, and pressure is reduced in the overlapping and side gap zone than Type-2 SAHT. At a 45° attack angle, the pressure difference across blade 1 is also reduced. Some vortical structure is also seen, which is less dominating during blade advancements for Type-2 SAHT.

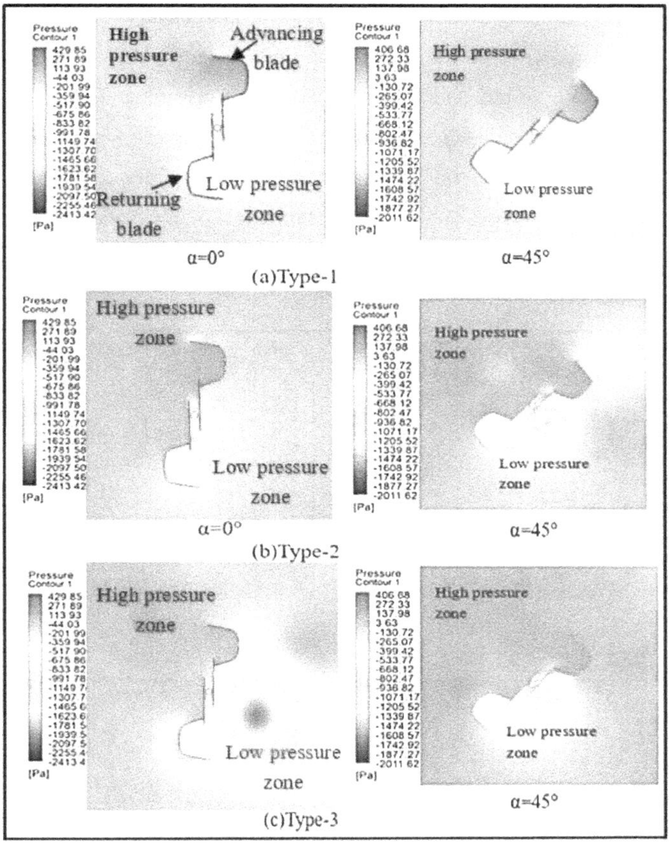

Fig.5.24 Static pressure signatures for (a)Type-1, (b)Type-2 and (c) Type-3 SAHT's at 0° and 45° attack angle for a flow stream-speed 0.5m/s

5.4.5 Vorticity characteristics

Figure 5.25 (a-c) displays the vorticity signatures of Type-1, 2, and 3 SAHTs with an attack angle of 0° and 45° with a flow stream-speed of 0.5 m/s. There is an observed vorticity range of 0 to 30 (s-1), and vorticity lowers the turbine's efficiency by preventing flow from exerting thrust force on the blades. The lower vorticity improves the turbine's performance by reducing the barrier for fluid to come into contact with the blades [8]. In Fig.5.25(a), at α=0° and α=45° for Type-1 SAHT, the high stall vortex zone adjacent to the tip of the advancing blade is observed. The high vortices patches are also observed near overlapping portions for both attack angles. But in Fig.5.25(b), there are low vortices for the same attack angle of 0° and 45° and small thin vortices are observed leaving both the blades in downstream, which indicates that the Type-2 SAHT blades experience a high-pressure effect since there is relatively less vortex region on their convex and concave faces. Low-pressure effect on the blades because of high-intensity deep stall vortex patches are located at pressure and suction sides for advancing and downstream-side blades similar to [9]. Again, in Fig.5.25(c), high vortices patches are formed on the downstream side and the concave side of the returning blade, which causes a negative pressure effect on the blades, reducing torque. Moreover, the Type-2 SAHT's overlapping and blade side gap contains smaller vortices than Type-1 and Type-3 SAHT.

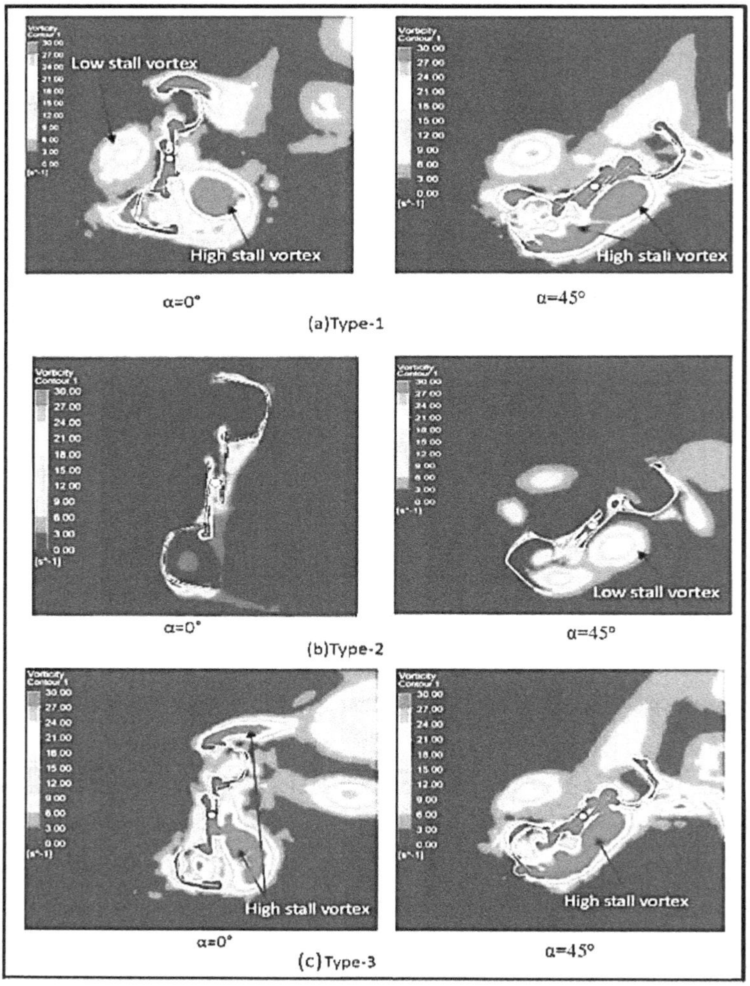

Fig.5.25 Vorticity signature of SAHT (a)Type-1, (b)Type-2 and (c) Type-3 SAHT's at α = 0° and 45° for a flow stream-speed 0.5m/s

5.4.6 Comparison of skin friction coefficients (C_f) among the turbine types

To determine the separation point at various attack angles, the skin friction coefficients (C_f) for Type-1, Type-2, and Type-3 SAHT are contrasted with regards to non-dimensionalised chord length at different attack angles of 4°,12° and 16° as represented in Fig.5.26 (a-c). In some previous research studies, it is found that stall occurs for attack angle in the range 4-18° [18], [19,20] and these angles are considered to determine the flow attachment on the blade surface from lower to higher attack angles. As seen from Fig.5.26 (a-c) for Type-2 SAHT, for all the three attack angles, the skin friction coefficient is higher and flow stays connected to the blade over the majority of the chord length positions, which is mostly due to effective flow control through blade curvature and overlapping and side gap flow. Moreover, compared to Type-1 or Type-3 SAHT, Type-2 SAHT tends to achieve the separation condition later, signifying effective blade fluid interactions.

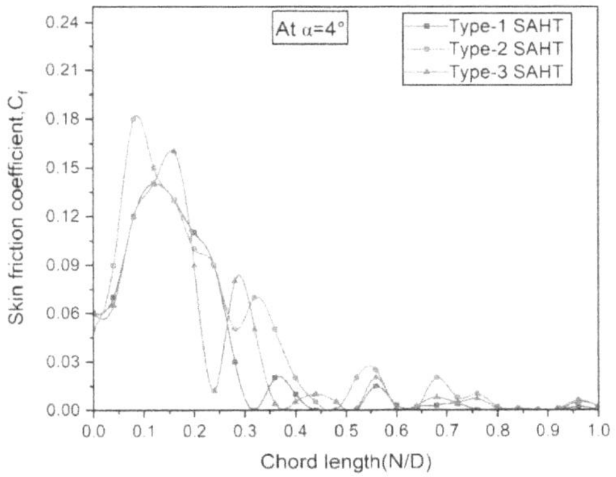

(a)

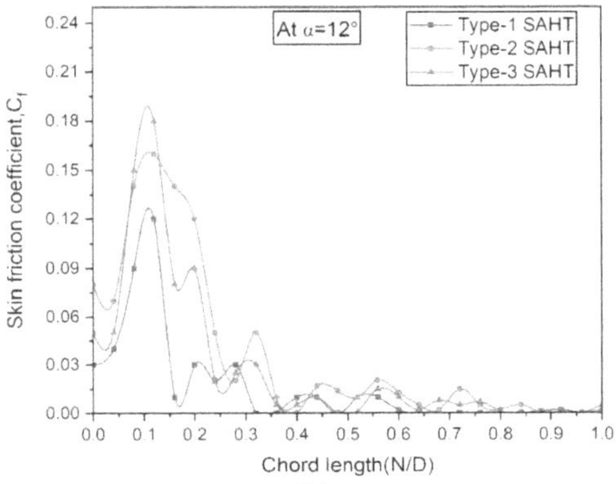

(b)

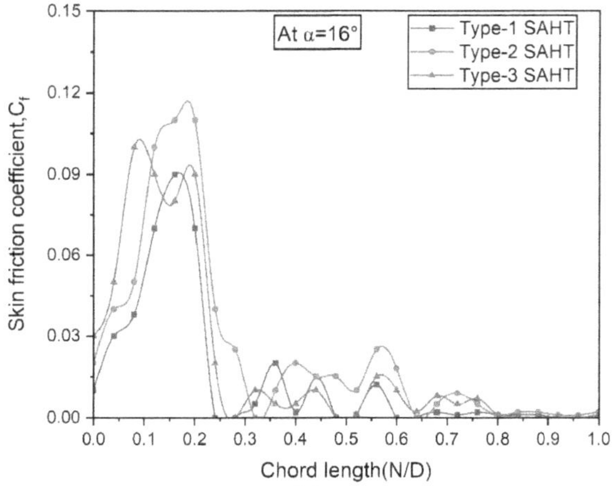

(c)

Fig. 5.26 C_f along chord length of optimal curved airfoil at 0.5 m/s flow stream-speed for (a) $\alpha = 4°$ (b) $\alpha = 12°$ (c) $\alpha = 16°$

5.4.7 Velocity characteristics

Velocity contour analysis can also be used to understand the flow field surrounding the turbines, i.e. Type-1, 2, and 3 SAHTs. The velocity contours at an attack angle of 0° are shown to understand the flow dynamics around turbines that operate at a *TSR* of 0.7. Figure 5.27(a-c) depicts a high-velocity region near the advancing blade and wake regions in blue patches downstream of Type-1, Type-2, and Type-3 SAHTs. In Type-1 and Type-3 SAHTs, i.e., Fig.5.27(a and c), the high-velocity region is low near the advancing blade pressure side, overlapping and blade side gap section, and the wake region is high at the concave side of the returning blade resulting in a reduction of overall performances. But in Fig.5.27(b), i.e., in Type-2 SAHT, it can be observed that there are higher velocity zones in the overlapping and blade side gap section, wider high-velocity region on the advancing blade's pressure side, and reduced wake region at the concave (suction) side of the returning blade, all these contributing to the better performance of Type-2 SAHT.

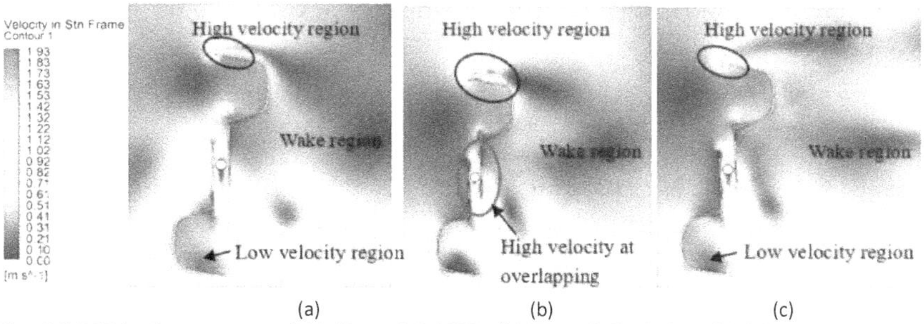

Fig. 5.27 Velocity contour of (a) Type-1 SAHT, (b) Type-2 SAHT and (c) Type-3 SAHT

5.4.8 Effect of flow stream-speed over torque for Type-2 SAHT

To verify the influence of flow stream-speed (0.4-0.6) m/s over torque (Fig.5.28) for the Type-2 SAHT, the CFD simulation is further performed for Type-2 SAHT, since the flow stream-speed is a controlled parameter and not a design parameter. At the beginning of flow time, i.e., from 0 to 2.2 s, the torque fluctuation is not uniform for all the flow stream-speeds, but after 2.2 s, the fluctuation becomes uniform. This is because the turbine initially takes time to accommodate the water thrust on the blades for rotation (transient phase), which results in non-uniform rotation, causing fluctuation of torque. In Fig.5.28, it is clearly noticed that the torque curve fluctuation at flow stream-speed 0.5 m/s is smaller than other cases (0.4 and 0.6) m/s and the average torque of 0.485 Nm is attained for 0.5 m/s flow-stream speed after flow time 2.2 s. The negative torque is lesser than at flow stream-speed of 0.4 m/s and 0.6 m/s from flow time 2.2 s onwards. Therefore, it can be stated that the flow stream-speed 0.5 m/s is the best flow condition for this design condition.

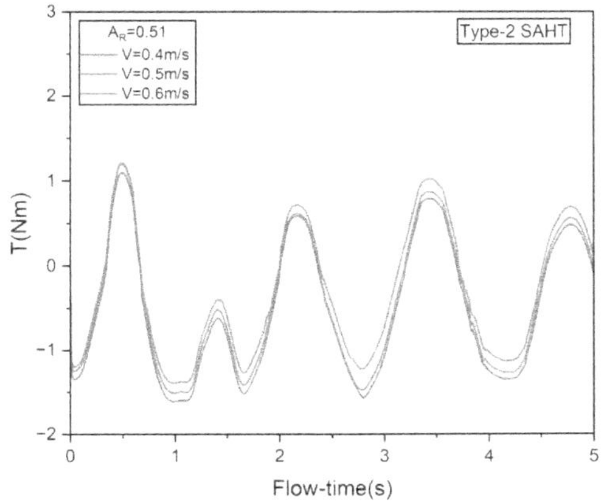

Fig.5.28 Flow stream-speed effect on torque (Nm) for Type-2 SAHT

5.4.9 Effect of flow stream-speed on tangential thrust coefficient for Type-2 SAHT at A_R=0.51

Figure 5.29 shows the turbine performance for the tangential thrust coefficient (Th_{coeff}) at flow stream-speeds 0.4, 0.5, and 0.6 m/s. Th_{coeff} is closely related to the turbine's torque and power output. The maximum value of the thrust coefficient is 0.89 at 0.5m/s flow stream-speed, which is obtained at an attack angle of 18°. Th_{coeff} values of 0.85 and 0.78 are obtained at 0.4 m/s and 0.6 m/s for the same attack angle. It is evident that when flow stream-speed rises, Th_{coeff} rises as well, indicating an increase in turbine performance. Interestingly, at the flow stream-speed of 0.5 m/s, the negative fluctuations in Th_{coeff} are lowest and it is also observed that when the flow stream-speed increases above 0.5m/s, tangential thrust coefficient decreases because the blade fails to accommodate higher fluid pressure due to increase of flow stream-speed and design incompatibility. Therefore, the SAHT can perform better at a flow stream-speed of 0.5 m/s with proper blade-fluid interactions and significantly prevent blade performance loss.

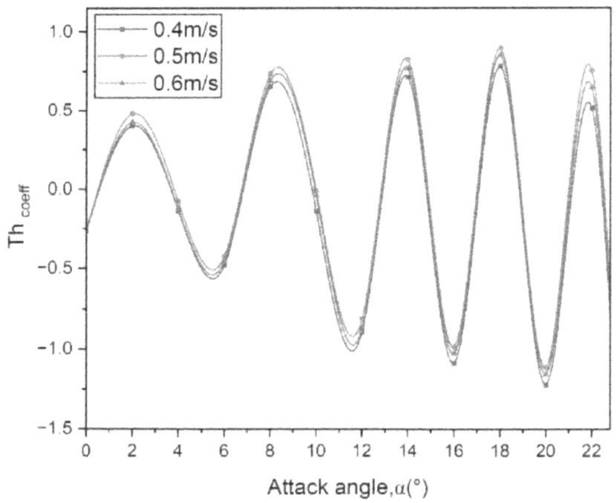

Fig.5.29 Flow stream-speed effect on Th_{coeff} for Type-2 SAHT at different attack angles

5.4.10 Effect of TSR on *Cp* and T_{coeff} at different flow stream-speeds for Type-2 SAHT

Hydrodynamic outcomes such as *Cp* and T_{coeff} with *TSR* (0.3, 0.4, 0.5, 0.6, 0.7, 0.9, and 1) for Type- 2 SAHT at 0.4, 0.5, and 0.6 m/s flow stream-speeds, as depicted in Fig. 5.30(a) and (b). Figures 5.30 (a) and (b) show that, for all flow stream-speeds, when *TSR* rises to 0.7, *Cp* and T_{coeff} increase, and when *TSR* rises more, they gradually decrease. In Fig.5.30(a), with the best flow stream-speed of 0.5 m/s, the SAHT functions at its maximum turbine *Cp* of 0.272. In Figure 5.30(b), as *TSR* grows, which shows that T_{coeff} increases until it reaches 0.68 at *TSR* 0.7, after which it declines. Thus, for a 0.5 m/s stream-speed, Type-2 SAHT functions better under all *TSR* values, peaking at 0.7 *TSR*.

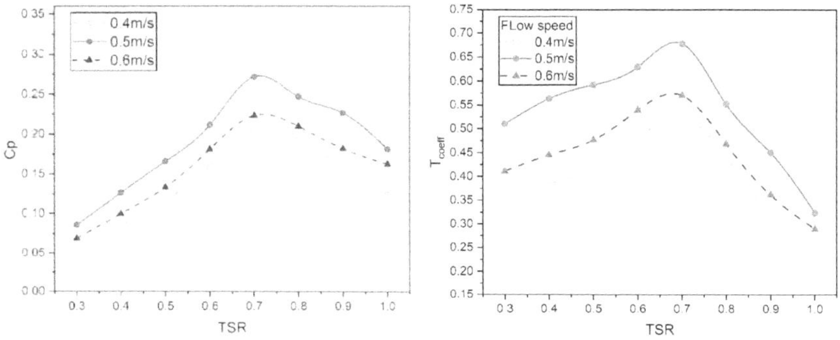

(a) (b)

Fig.5.30 Flow stream-speed effect on (a) *Cp* and *(b)* T_{coeff} for SAHT with respect to TSR (0.3-1)

Table 5.4 displays the highest water-power input (P_{in}) along with the associated power output (P_{out}) using *Cp*. In Table 5.4, it is seen that for an optimal flow stream-speed 0.5 m/s, the P_{out} per unit height for Type-2 SAHT is 0.476W, and the corresponding *Cp* is 0.272. Now, considering P_{out} =0.476 W for the investigated Type-2 model, the prototype power (P_P) can be calculated assuming a suitable scale ratio, say 27:1, using the relation from Eq.13(Chapter 3 Section 3.5). Using this equation if the model turbine diameter is 0.334 m, then the diameter of the prototype (D_P) comes out as 9 m. Using the optimal aspect ratio of 0.51, the prototype turbine height (H_P) comes out as 4.5 m. Therefore, the Pico-scale prototype power (P_P) obtained from Eq.6(Chapter 3 Section 3.5) is 2.5 kW, which can fulfil the power requirements of 15 households (approx.) considering 150W individual household requirements, generally found in remote locations near perennial river streams, canals, etc., [18]. Such a prototype SAHT can be thought of installing on the downward slope from the riverbank to its bed or in similar designs of low discharge channels, which can produce water-power in a sustainable manner.

Table 5.4 Details of water-power input (P_{in}), power output (P_{out}), and *Cp*

V (m/s)	P_{in} (W)	P_{out} (W)	*Cp*

0.4	0.896	0.183	0.204
0.5	1.75	0.476	0.272
0.6	3.024	0.675	0.223

5.4.11 Influence of Turbulence intensity (T.I.) on Type-2 SAHT performances

Figures 5.31 (a-g) illustrate how the turbulence intensity *(T.I.)* affects lift coefficient (C_L) and drag coefficient (C_D) for Type-2 SAHT at different attack angles (0 to 22) ° for flow stream-speeds 0.4 m/s, 0.5 m/s and 0.6 m/s, respectively. Figure 5.31 (a) illustrates that the C_L is minimum at lower turbulence intensity, and C_L increases with increasing turbulence intensity up to 10% and then decreases as *T.I.* increases further at these three-flow stream-speeds. It is also seen that at higher *T.I.* of 12%, C_L is decreased, which can be due to the generation of vortices [8] near the blades, which causes energy losses at the blade and reduces lift coefficients [4, 21]. In Fig.5.31(a), (b), and (c), it is observed that C_L rises gradually up to an attack angle between (16 to 18) °, and when the attack angle increases further, C_L subsequently drops. Type-2 SAHT demonstrated its best results at an attack angle of 18 ° with a higher C_L value of 0.95. The optimal curve profile used is more stable [7] due to the smallest fall in C_L for Type-2 SAHT. Moreover, the stall angle of Type-2 SAHT has shifted to a higher attack angle (18)° than Type-1 (14)° and Type-3(16)° SAHTs at flow stream- speed 0.5 m/s compared to the other flow conditions, signifying effective flow control at flow stream-speed 0.5 m/s. Furthermore, it is also seen that as *T.I.* rises gradually, C_D also increases for all three SAHTs, and it is higher at 12 % T.I., as shown in Fig.5.31 (d-f). A similar C_D pattern was also observed in [17, 22].

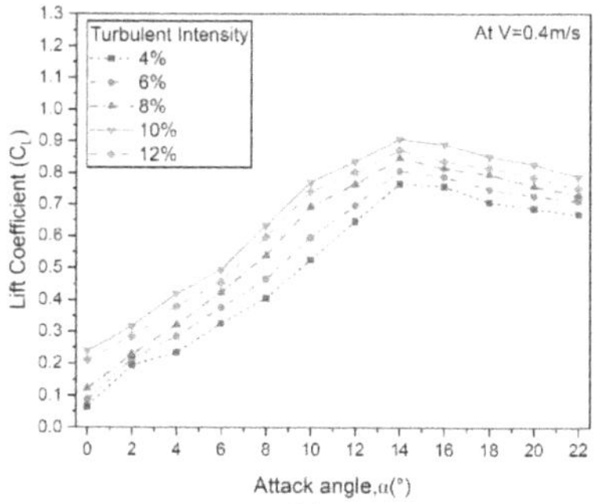

(a)

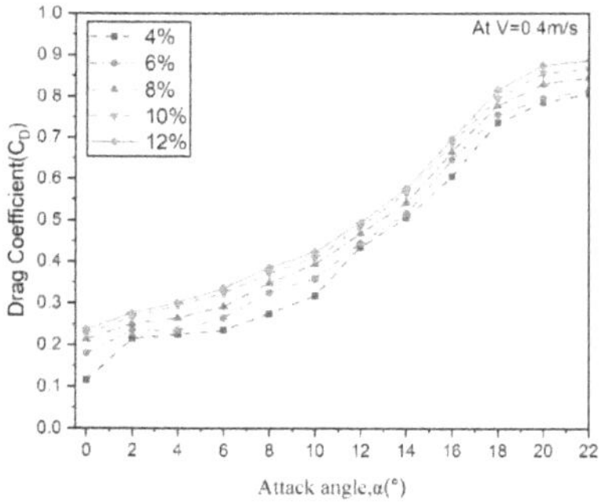

(d)

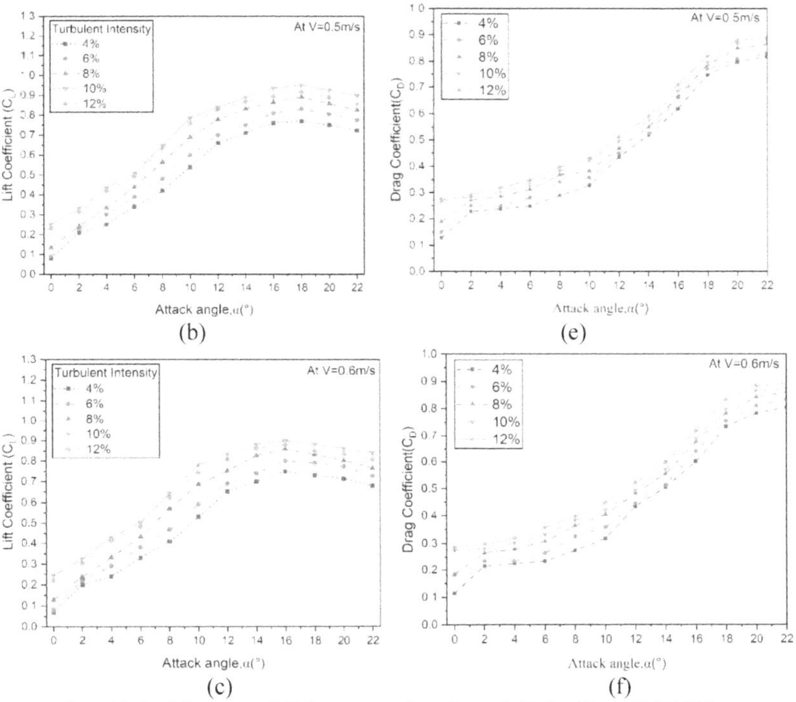

Fig.5.31 (a-f) Impact of *T.I.* representing C_L and C_D for Type-2 SAHT

5.4.12 Effect of overlapping over velocity vectors around the Type-2 SAHT

The velocity vectors across the Type-2 SAHT blades are shown (Fig. 5.32) at $\alpha = 0$ ° at flow stream- speed 0.5 m/s, 0.7 TSR, and different overlapping (φ) (20, 25, 30, and 35) % to highlight the flow pattern across the turbine. It can be seen from Fig.5.32(a) that for overlapping 20%, increased magnitude of velocity vectors can be found across the blades, and more importantly, higher velocity distributions can be found on the suction surfaces of blade-1 and blade-2. In addition to this, overlapping and gap width flow is much developed, having increased velocity magnitudes at 20% overlapping conditions, for which this flow control impinges more thrust on the suction surface of the returning blade. Further, Coanda-like velocity vectors proceed towards the returning blade (blade 2) from the downstream side adjoining the suction surface of the advancing blade (blade 1). Furthermore, as overlapping is increased to 25% and higher (Fig.5.32b-d), the velocity field suffers from vortices effects (Fig.5.32b-d), especially on the downstream of the bottom side blade, which results from the reduced flow control, causing loss of momentum.

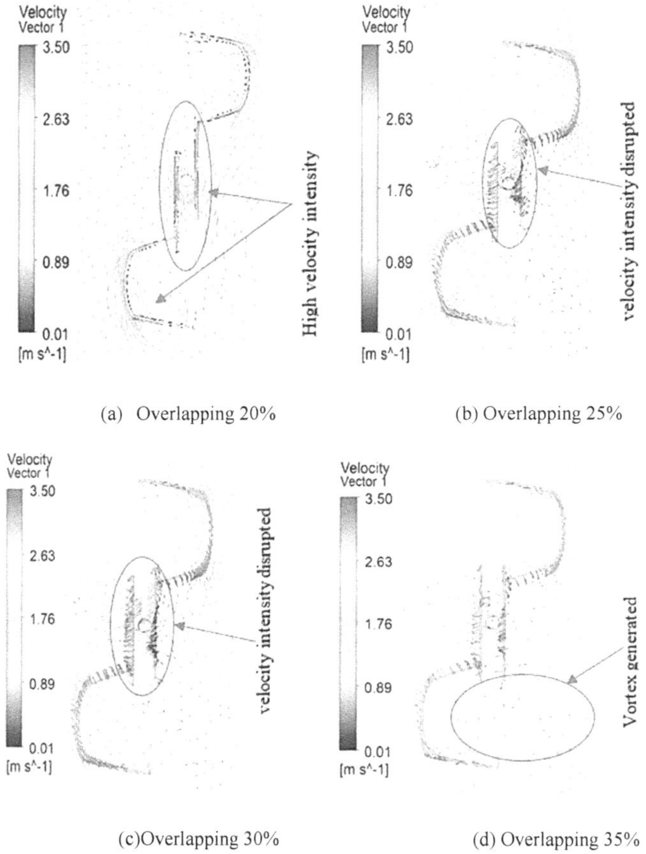

Fig. 5.32 Velocity vectors signatures for different overlapping (a) 20%, (b)25%, (c)30%, and (d)35% at $\alpha=0°$

5.4.13 Comparison of velocity vectors for Type-1,2 and 3 HSTSs

Seeing the improved performance at 20% overlapping in the previous section, it is desirable to compare the flow physics in terms of velocity vectors of the three turbine types having the combination of three individual arc types and 20% overlapping and fixed gap width condition. Compared to Type-1 and Type-3 SAHT, Type-2 SAHT (Fig.5.33b) has a higher concentration of velocity vectors on the pressure side of the advancing stage blade (blade 1), which is contributed by the effective arc curvature flow control by the Type-2 turbine. Overlapping and gap width flow control is also seen effective for Type-2, like at $\alpha=0°$ condition. These features improve the pressure, velocity, and other aerodynamic signatures of the Type-2 turbine blades, as seen earlier.

By comparing Fig.5.33(b) with Fig.5.33(a) or Fig.5.33(c), the offset (i.e., the gap between the inner edges of the curve section and straight blade section) flow control is also seen for Type 2 turbine by the presence of some velocity vectors there, whereas for Type-1 and 3, hardly any concentration of vectors can be seen. It can also be observed that the velocity vectors near the convex side (i.e., suction surface) of the advancing stage blade (blade 1) impinges sharply on the back side of its straight blade section and then proceed towards the concave section of the returning blade (blade 2) to produce extra thrust to that blade through this flow control from the downstream side of arc curvature. This is as good as emulating Coanda-like vectors migrating towards the returning blade, as also observed in the previous result. In this process of vector migration, some vortical regions are also formed. However, this region is away from the blade contacts at the straight blade section, whereas the same is attached to the straight blade section for Type-1 and Type-3 SAHTs. Further, small vortical regions are seen in various locations adjoining the blades for Type-I and Type-3 SAHTs. Therefore, overall, there are various effective flow controls for Type-2 SAHT, in fact, five in number, i.e., advancing blade's arc curvature flow control (both pressure and suction sides), overlapping and gap width flow control, offset flow control, Coanda-vector like flow control, and vortical flow control. Because of such effective blade-fluid flow controls and interactions, the optimal Type-2 SAHT (as per Taguchi optimization) with overlapping 20% produces higher hydrodynamic performances for the present design of SAHT.

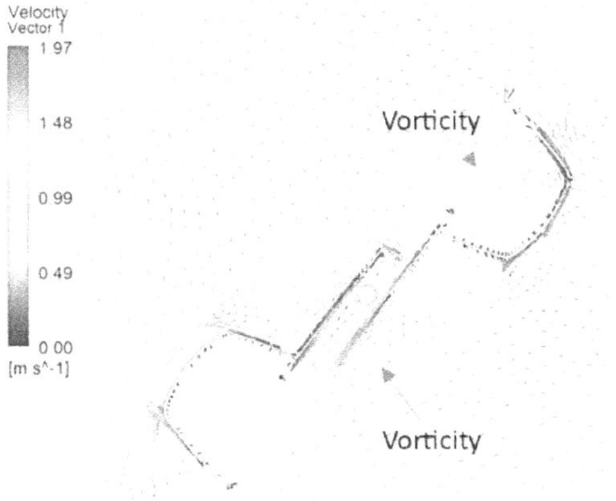

(a)Type-1 SAHT

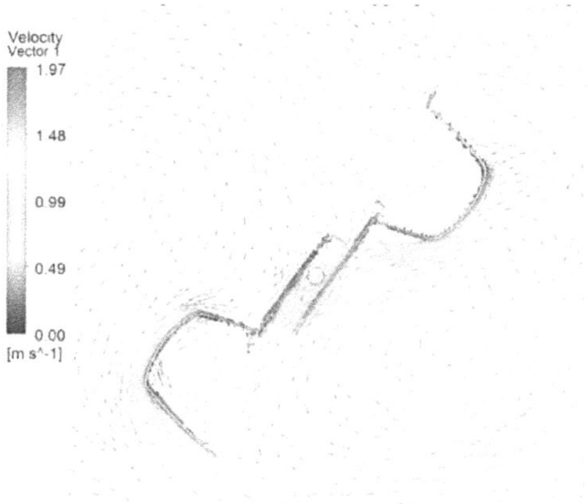

(b) Type-2 SAHT

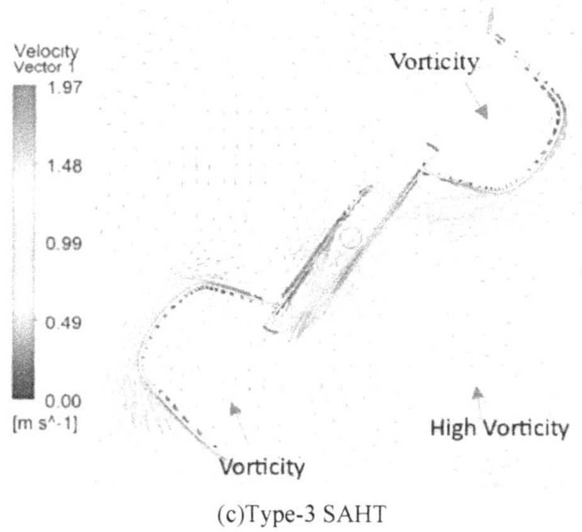

(c)Type-3 SAHT

Fig. 5.33(a-c) Velocity vectors signatures for 20% overlapping at α=45° for Type-1,2 and 3 SAHTs

5.4.14 Effect of blade thickness for Type-2 SAHT at optimal overlapping

The influence of blade thickness (t) of (1.5, 2, 2.5, and 3) mm are examined for Type-2 SAHT at overlapping of 20% to see the improvement in coefficient of performances (*Cp*). The *Cp* with respect to *TSR* for these four blade thickness is shown in Fig.5.34. It can be observed that the *Cp* has reached a maximum value of 0.278 with a maximum percentage increase in *Cp* by 2.20% for t=2mm at 0.7 *TSR* and when the thickness is increased further performance deteriorates. The reason behind this is the blade finds it difficult to rotate at this flow stream-speed, which is probably due to the unproductive increase of overall inertia of the blades due to the increase in thickness. It is also observed that as the *TSR* rises beyond 0.7, performance reduces. The reason behind this is the blade fails to accommodate the higher flow stream-speed at the tip of the blade during rotation. Therefore, it can be concluded that the blade thickness of 2 mm is suitable for Type-2 SAHT at an overlapping of 20% for improving the performance of the turbine. Figure 5.35 demonstrates the variation of T_{coeff} with respect to *TSR* for t=1.5-3 mm. It is also observed that T_{coeff} increases up to 0.702 with maximum percentage enhancement by 3.23% for t=2 mm with *TSR* 0.7 and subsequently falls as *TSR* rises even higher. Consequently, at a flow stream-speed of 0.5 m/s, SAHT functions better under all *TSR* values, peaking at 0.7 *TSR*.

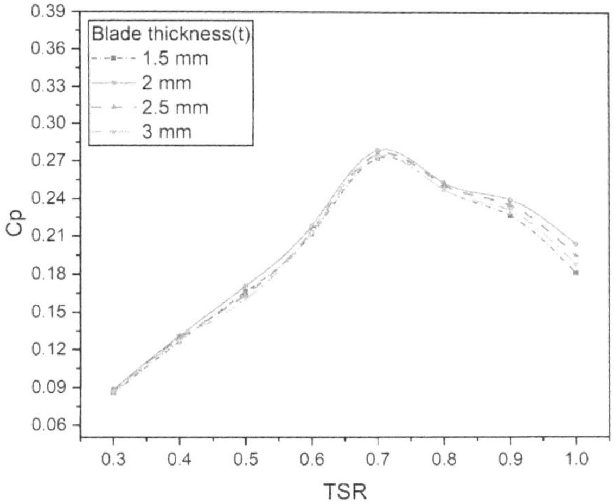

Fig. 5.34 *Cp* with respect to *TSR* for different blade thickness (1.5-3) mm at 20% overlapping for a flow stream-speed of 0.5m/s

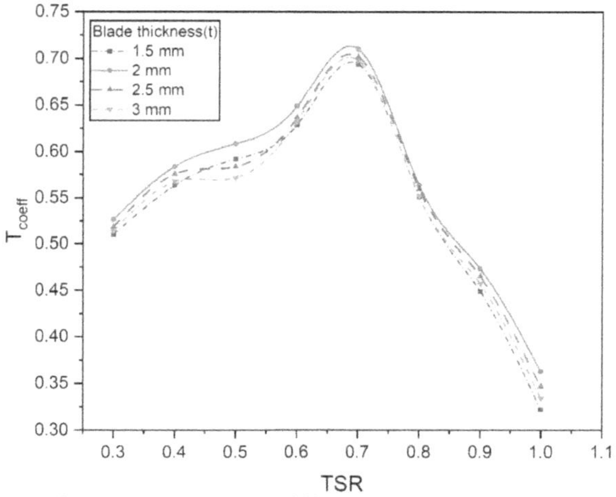

Fig. 5.35 T_{coeff} with respect to *TSR* for different blade thickness (1.5-3) mm at 20% overlapping for a flow stream-speed of 0.5m/s

5.5 Comparison with the existing literature

As seen in Table 5.5, the *Cp* result of Type-2 SAHT is compared with the studies of Basumatary et al. [16] and Khan et al. [9], with almost similar flow stream-speed conditions (0.3-0.5) m/s, *TSR* range (0.4-1.1) and aspect ratio (0.5-0.7). The highest *Cp* for SAHT (0.278) is higher than [16] (i.e., 0.268) and higher than [9] (0.15). When compared to [16] and [9], the percentage *Cp* enhancements of the present SAHT are 3.73% and 85.33% over the works of [16] and [9].

Table 5.5 Comparison with the existing literature

Reference/Section no.	Year	Type of airfoil based design	Turbine Size D_T (mm)	Flow-speed(m/s)	*CP*
Basumatary et al.[16]	2021	NACA 0018 airfoil-straight bladed	250	0.5	0.268
Khan et al.[9]	2022	S1048 airfoil-straight bladed	330	0.38	0.15
Model-4	-	NACA 63415 airfoil-straight bladed	334.62	0.4	0.36
Type-3	-	NACA 63415 Airfoil-straight bladed With arc Curvature radius	33.4.62	0.5	0.278

5.6 Summary of the results

In the current work, the hybrid Savonius hydrokinetic turbine (SAHT) i.e., Model-4 under which (Type-1 to Type-3) were designed, and analyzed to improve their performances. The SAHT is designed with blade arc curvature, overlapping, gap width, and offset at best arc radius, aspect ratio, and flow speeds. The different hydrodynamic parameters, along with flow physics around the SAHTs' are studied using 2D CFD simulation. The following results can be drawn from the study:

- The Type-2 SAHT exhibits lower torque fluctuations and higher average torque at the best A_R = 0.51, with φ = 20 % and ε = 6.14 %. The blade thickness, t = 2 mm, has further raised its performance.

- The Type-2 SAHT shows the highest C_L and C_D of 0.95 and 0.9 obtained at a flow stream-speed of 0.5m/s for 10% turbulence intensity. The R_2 =15 mm, regulates effective flow controls on both the advancing and returning blades, causing effective pressure, velocity, and hydrodynamic performance signatures. Further, R_2 =15 mm ensures Coanda-vectors like flow control to impact additional thrust on the returning blade.

- Vortical flow structures across the Type-2 SAHT are less developed and are not attached to the blades of the Type-2 turbine, meaning better stall regulation for this optimal turbine type. Further flow controls through overlapping and gap width flows, as well as offset flows, depict positive blade fluid interactions, raising the hydrodynamic performances of Type-2 SAHT.

- The overlapping of 20% enhances the overall performance of the Type-2 SAHT by allowing maximum flow through overlapping and causing more fluid interaction with the blades.

- The highest Cp obtained for the optimal SAHT (with ψ = 20 %, R_2 = 15 mm, A_R = 0.51, and t =2 mm) was 0.278 at 0.7 TSR and flow stream-speed 0.5 m/s. The Type-2 SAHT exhibits the highest T_{coeff} of 0.702 for the optimal design for a flow stream- speed of 0.5m/s. The Cp and T_{coeff} of SAHT were improved by 2.20% and 3.23%, respectively, with the best blade thickness of 2 mm.

- Considering A_R = 0.51, the prototype power of 2.5 kW can be obtained for flow stream-speed of 0.5 m/s. Such a prototype turbine can be installed in the downward slope from the riverbank to its bed or in similar designs of low discharge channels, which can produce water power in a sustainable manner.

5.7 Experimental analysis of the hybrid SAHTs

In this section, the Model-4(section 5.3 size has been reduced by scaling down to form Model-5 and the performance parameters such as T (Nm), P (W), Cp and T_{Coeff} are determined at different load conditions (250-2000) g and results are analyzed and discussed in detail. The scale ratio in terms of power output (W) for Model-5 is determined. Further, the performance in terms of Cp is investigated by performing CFD analysis keeping the same design parameters of Model-5. Finally, to understand the flow around *SAHT*, the flow structure phenomena are also discussed and corroborated with the experimental analysis.

5.7.1 Investigations of different aspect ratios (A_r) of (a)0.22, (b)0.26 and (c) 0.29 on SAHT

The performance of the turbine is greatly influenced by its aspect ratio [23]. The different aspect ratios such as 0.22, 0.26 and 0.29 are investigated without blade gap width and overlapping at the beginning and plotted in Fig. 5.36(a-c). The performance in terms of Cp are compared at different stream-speeds (0.6-0.9) m/s. In Fig.5.36 (a-c), the Cp pattern for all stream-speed conditions rises continuously to a maximum load of 1500 g and when the load increases, the Cp decreases. The turbine can still perform well up to maximum load 1500g, and as the load increases beyond 1500g, the turbine performance is reduced due to the decrease in turbine rpm due to heavier loads on the turbine blades [24]. The maximum Cp obtained with A_r = 0.22, 0.26, and 0.29 are 0.045, 0.051, and 0.041, respectively at load 1500 g for stream-speed of 0.8m/s as shown in Fig.5.36 (a-c). This is because, for maintaining 0.8 m/s stream speed through the channel, the turbine of A_r = 0.26, is possible to be submerged 1cm below the top layer of the water stream and at this position the turbine can perform better due to maximum fluid interactions with the blades. But as the aspect ratio increases above 0.26, some portion of the blade remains to be above water top layer, which causes improper fluid interactions with the turbine blades resulted in energy loss in the turbine blades. At lower stream-speed of 0.6 m/s and 0.7 m/s, the Cp is lower than at 0.8m/s, but when the stream-speed increases beyond 0.8 m/s, Cp again deteriorates. At higher stream-speeds, the blade finds difficult to accommodate the higher thrust on the blades, which causes loss of volumetric discharge and blade tip speed losses and due to this, the performance reduces [25]. Therefore, due to all these factors, the aspect ratio of 0.26 is found to be better for the hybrid turbine design at 0.8m/s stream-speed in the current experimental set-up.

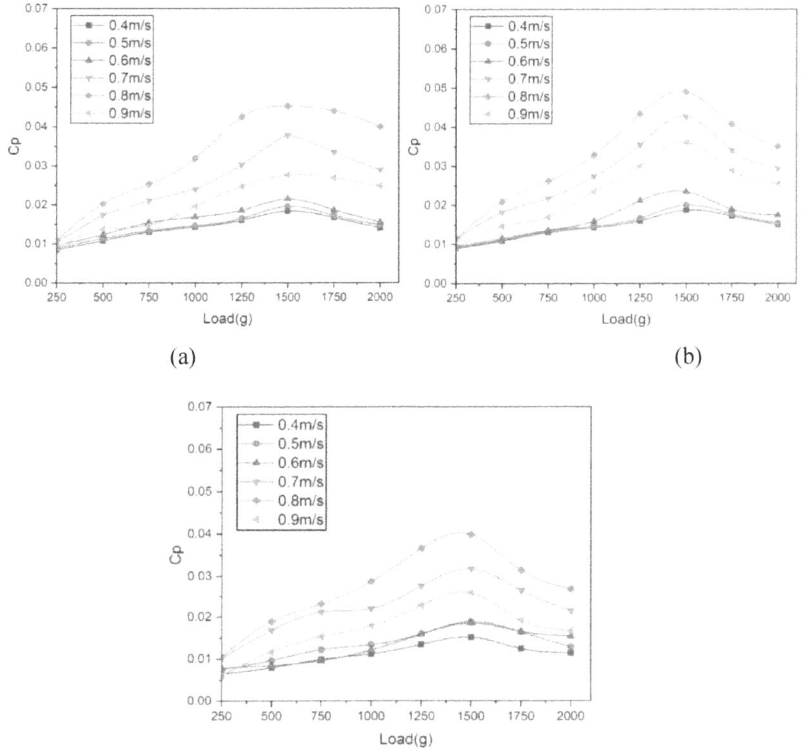

(a) (b) (c)

Fig.5.36 *Cp* of turbine with aspect ratios (a) 0.22, (b) 0.26 and (c) 0.29 for flow speeds (0.4-0.9) m/s

5.7.2 Influence of blade gap width ratio (ε) of (4.16-7.10) % on *Cp* for stream-speed (0.6-0.9) m/s

The blade gap width ratio (ε) generally increases with the distance between the turbine blades, indicating smoother water flow passage between the turbine blades, which can improve turbine performance [26]. In this case, the flow stream speed range has been investigated from 0.6-0.9 m/s since the performance of SAHT deteriorates, when flow speed decreases below 0.6m/s (section 5.7.1). Figure 5.37 (a-d) shows the effect of stream-speeds on *Cp* for blade gap width ratio (ε) of 4.16%, 6.14%, 6.70%, and 7.10% for best Ar = 0.26 at stream speeds (0.6-0.9) m/s. It is observed that *Cp* is maximum up to 1500g brake load and 0.8 m/s stream speed in all blade gap width ratio cases. In Fig. 5.37(a), the maximum *Cp* of 0.130 is obtained for blade gap width ratio of 4.16% at 0.8 m/s stream-speed. In Fig.5.36 (c) and (d), the maximum *Cp* of 0.149 and 0.126 are obtained at ε = 6.70% and 7.10%. But in Fig. 5.37(b), at ε =6.14%, the turbine can generate maximum *Cp* of 0.173 which is higher than in Fig.5.37 (a), (c) and (d). This is because at blade side gap higher than 6.14%, the gap between the blades becomes high and as a result, the water bypasses through the higher side gaps without interacting with the turbine blades causing reduction in turbine performances [26, 27].

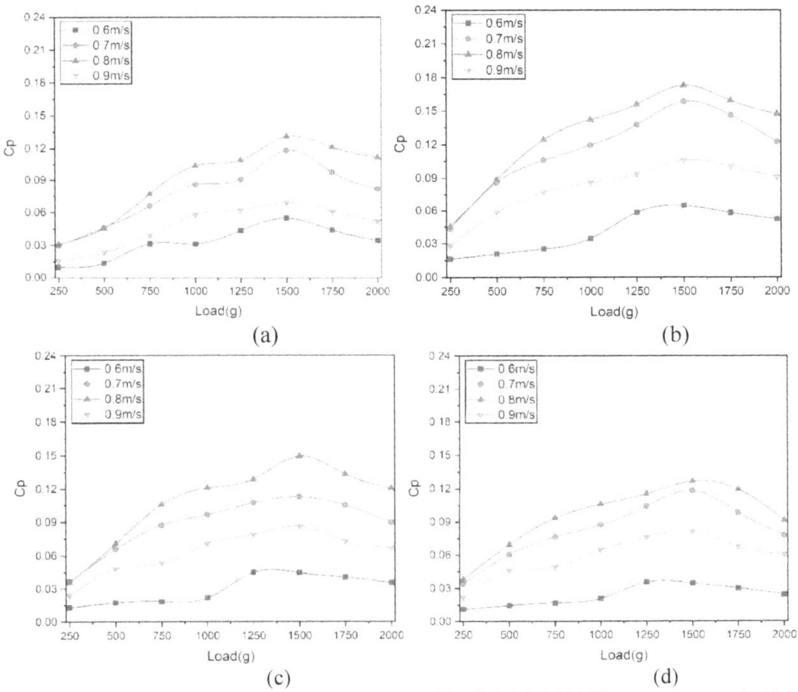

Fig.5.37 Comparison of *Cp* at blade gap width ratio of (4.16-7.10) % at stream-speeds (0.6-0.9) m/s

5.7.3 Influence of overlapping (15- 30) % on Power output (W) at best gap width ratio, ε=6.14%

The effect of overlapping on turbine performances is determined keeping the aspect ratio fixed at 0.26 and it is same for the rest of the experimental investigations. The influence of overlapping, φ (15-30) % on power output, P (W) at ε = 6.14% and A_r=0.26 are depicted in Fig.5.38 (a-d). The overlapping ratio of (15, 20, 25, and 30) % are investigated as it is known that overlapping increases the performance of Savonius turbines [25, 28, 29]. The influence of overlapping has been investigated by increasing flow speed up to 1.0m/s, which is the maximum attainable flow speed in the recirculating water channel. As seen in Fig.5.38 (a-d), the power output pattern at stream-speeds (0.6-1.0) m/s are also similar as in case of *Cp* (section 5.7.2), showing the fewest disparities among the variations for overlapping of (15-30) % respectively. The maximum power of 0.227 W, 0.327 W, 0.209 W, and 0.154 W are obtained at 1500 g load for overlapping of 15%, 20%, 25%, and 30% respectively. Since, power is directly proportional to torque (Eq.4), and at 20% overlapping, the turbine rotates at higher angular velocity of 3.41 rad/s for 0.8 m/s stream-speed. Therefore, the *SAHT* with φ= 20% and optimal ε = 6.14% can generates highest power at optimal low stream-speed 0.8 m/s. The reason why turbine power decreases with the increase in stream-speed above 0.8 m/s at 20 % overlapping is also further discussed by investigating torque variation at same stream-speeds (section 5.7.4).

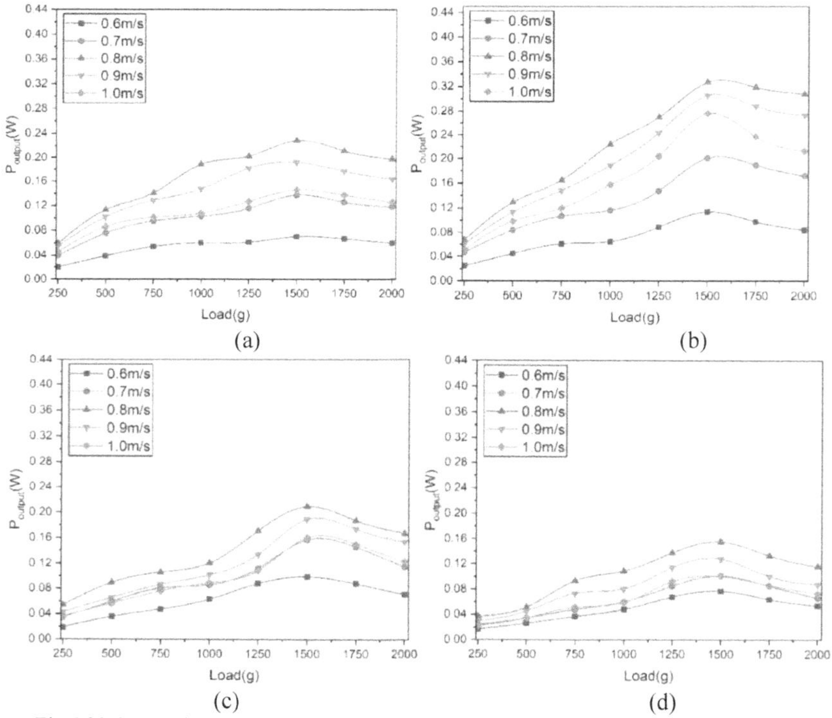

Fig.5.38 Comparison of output power (W) for ε = 6.14% with overlapping of (15- 30) % and stream-speeds (0.6-1.0) m/s

5.7.4 Influence of overlapping (15-30) % over Torque (Nm) at ε=6.14%

The different overlapping of (15, 20, 25, and 30) % at ε=6.14% are studied to determine the torque variation on *SAHT* at stream-speeds (0.6-1.0) m/s represented in Fig.5.39 (a-d). It is observed that, the torque is maximum at 1500 g load with 0.8 m/s stream speed for all overlapping cases. In Fig.5.39(a), the maximum torque of 0.079 Nm is obtained for 15% overlapping at 0.8 m/s stream-speed and the same decreases as the stream-speed increases up to 1.0 m/s. In Fig.5.39(c) and (d), the maximum torque of 0.075 Nm and 0.056 Nm are obtained at 25% and 30% overlapping. But in Fig. 5.39(b), the turbine can generate maximum torque of 0.096 Nm at 20% overlapping for the same stream-speed, which is higher than in Fig.5.39 (a), (c) and (d). This is probably because as the overlapping increases beyond 20%, the water escapes through the overlapping gap [3,27], which weakens the pressure effect on the turbine blades. This results in the loss of volumetric discharge in overlapping section, resulting in the low torque generation in the turbine [3, 27, 30].

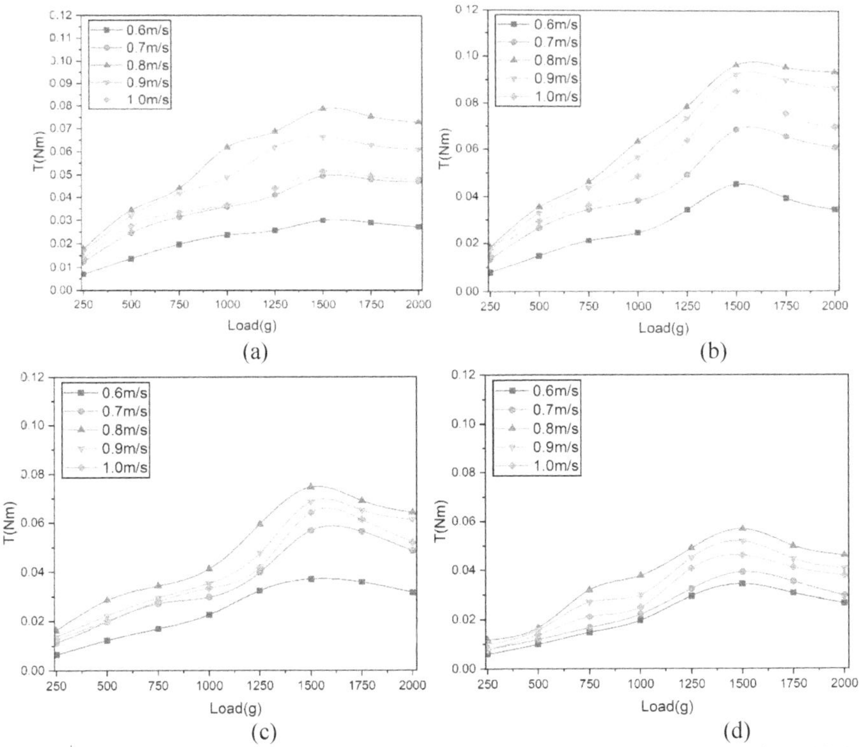

Fig.5.39(a-d) Comparison of torque (Nm) for best ε=6.14% with overlapping of (15- 30) % and stream-speeds (0.6-1.0) m/s

5.7.5 Influence of overlapping (15- 30) % on *Cp* at best ε=6.14%

Figure 5.40(a-d) shows the effect of overlapping (15-30) % on *Cp* at gap width ratio of 6.14% for stream-speeds (0.6-1.0) m/s. It is again observed that, the *Cp* is maximum up to 1500 g load and 0.8 m/s stream speed in all overlapping design conditions. In Fig.5.40 (a), the maximum *Cp* of 0.178 is obtained for 15% overlapping at 0.8m/s stream-speed. In Fig.5.40(c) and (d), the maximum *Cp* of 0.197 and 0.121 are obtained at overlapping of 25% and 30%. But in Fig. 5.40(b), at 20% overlapping, the turbine can produce maximum *Cp* of 0.256 which is higher than in Fig.5.40 (a), (c) and (d). At a stream-speed of 1.0 m/s and 30% overlap, the lowest *Cp* is 0.041. According to modified turbine size and swept area, the current blades have the capacity to support highest stream-speed till 0.8 m/s, at 20% overlap, and loads up to 1500 g, enabling water to generate the most desirable thrust on the blade.

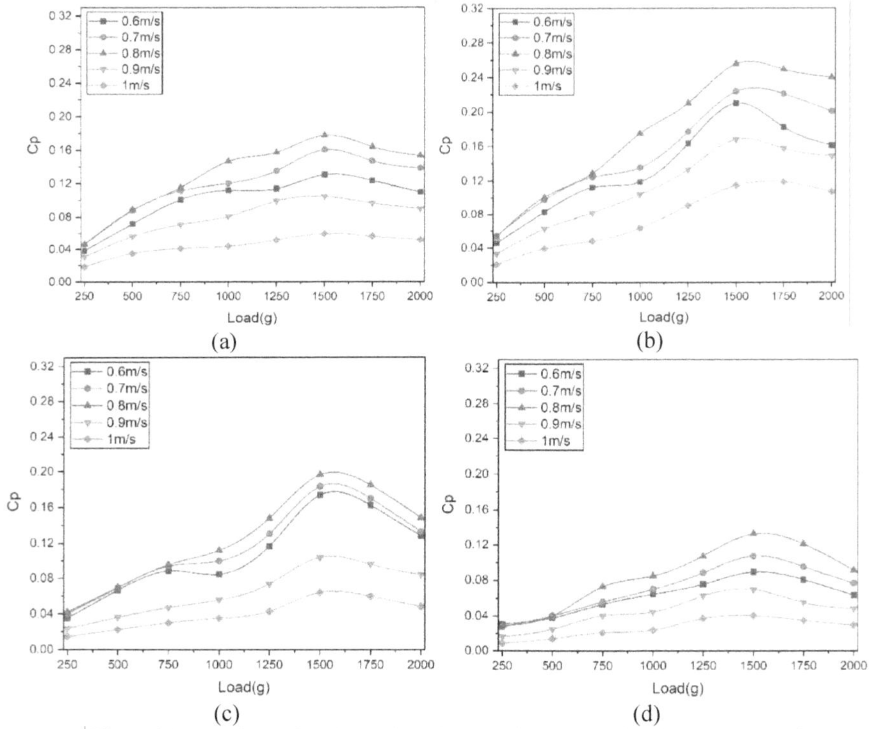

Fig.5.40 Comparison of *Cp* for optimal β_{SB}=6.14% with overlapping of (a)15%, (b)20%, (c)25% and (d)30 % for stream-speeds (0.6-1.0) m/s

Therefore, when the results are thoroughly examined above, the advantages of keeping 6.14% gap width, overlapping of 20%, and aspect ratio 0.26, the hybrid turbine, which is named as Model-5 shows better hydrodynamic performances up to 1500 g load for 0.8 m/s stream-speed. To highlight this point, Table 5.6 summarizes the values of *P* (W), *T* (Nm), and *Cp* for a load of 1500 g with optimal ε=6.14% and φ=20% in relation to the stream-speeds that were examined (0.6, 0.7, 0.8, 0.9 and 1.0) m/s.

Table 5.6 Experimental results at ε=6.14%, φ=20% and A_r=0.26 at 1500g load

Stream-speeds (m/s)	T (Nm)	P (W)	Cp
0.6	0.045	0.113	0.210
0.7	0.069	0.201	0.235
0.8	0.096	0.327	0.256
0.9	0.094	0.305	0.168
1	0.091	0.275	0.113

5.7.6 Effect of stream-speeds (0.6-1.0) m/s on T_{Coeff} at 1500g load

Figure 5.41 shows the torque coefficient (T_{Coeff}) fluctuations in *SAHT* at low speeds of (0.6-1.0) m/s while operating at the ideal brake load of 1500 g. It shows that the T_{Coeff} gradually rises from 0.372 to 0.445 and subsequently decreases as the stream-speed increases. The highest T_{Coeff} (0.445) is attained using best 20% overlapping, blade gap width ratio 6.14% and a stream-speed of 0.8 m/s. This is because the blades fail to accommodate the excessive water pressure due to increase of stream-speed greater than 0.8m/s, which causes pressure loss at the blade tips, which lowers torque and causes a loss of volumetric discharge. The present turbine blades' size, projected area, and offset 15 mm can support highest flow rates till 0.8 m/s speed condition, allowing water to apply maximum thrust on the blades.

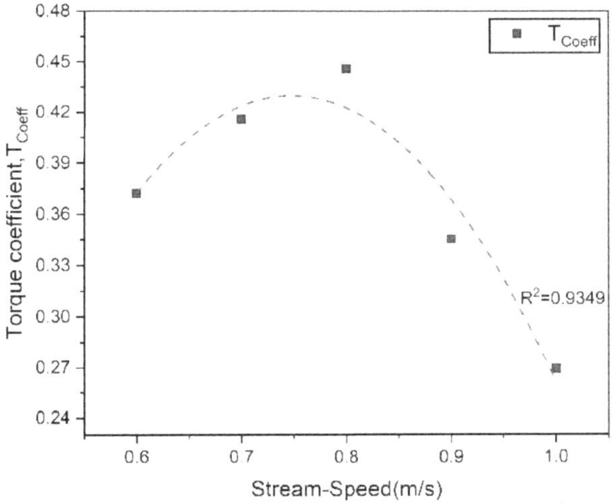

Fig.5.41 Maximum T_{Coeff} comparison for water currents of (0.6-1.0) m/s at 1500g load

5.7.7 Influence of TSR on maximum *Cp* at 1500 g load for (0.6-1.0) m/s stream-speeds

Figure 5.42 shows the influence of *TSR* over Max. *Cp* on *SAHT* at different stream-speeds (0.6-1.0) m/s for optimal blade gap width ratio 6.14% and overlapping ratio 20% at 1500 g load. It exhibits the maximum *Cp* that may be attained under different operating conditions, signifying the SAHT's final performance characteristics. In Fig.5.42, the *Cp* rises to maximum of 0.256 at 0.574 *TSR*, which is obtained at 0.8m/s stream-speed and then it decreases as the stream-speed increases above 0.8m/s. The lowest *Cp* obtained is 0.114 at 0.424 *TSR* for 1.0 m/s stream-speed. This is because the turbine blades' size, projected area, blade side gap, and overlapping enable water to exert maximum desirable force over blades by accepting maximum stream-speed up to 0.8 m/s at 0.574 *TSR* for 1500 g load. However, the stream-speed above 0.8 m/s, causes tip loss and the water leaves turbine without affecting the blades, which reduces performances of the turbine.

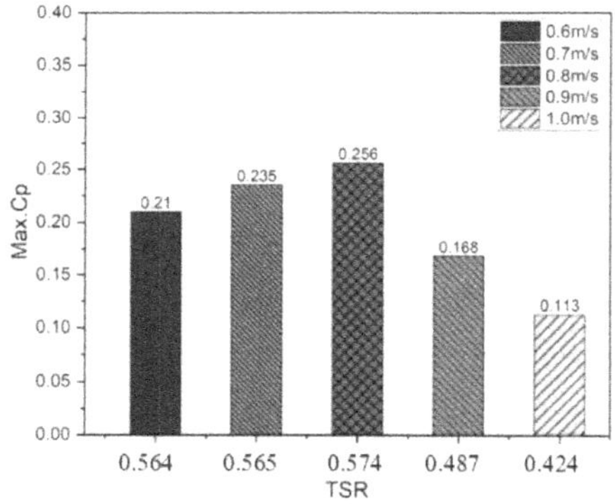

Fig.5.42 Effect of TSR on torque (Nm) for *SAHT* at a stream-speed of (0.6-1.0) m/s

5.7.8 Influence of *TSR* on T_{coeff} for overlapping and blade side gap ratio 20% and 6.14% at a stream-speed of 0.8 m/s

Figure 5.43 displays the torque coefficient (T_{Coeff}) change in the SAHT when it is running at the ideal load (1500 g) with respect to *TSR* for a stream-speed of 0.8 m/s. As RPM decreases, *TSR* also decreases as well due to gradual increase of load on the turbine. Though the RPM and *TSR* decreases, but the T_{Coeff} increases gradually up to 0.57 *TSR*, and then decreases as *TSR* decreases further. This is probably due to the reduction in the blade tip speed loss up to 0.57 *TSR*, since the current turbine blades can accommodate *TSR* of 0.57 and a maximum stream-speed of 0.8 m/s, enabling the water to apply maximum pressure on the blades. Therefore, at *TSR* 0.57, 20% overlapping ratio, blade side gap ratio 6.14%, and a stream-speed (0.8 m/s), the highest T_{coeff} (0.445) is attained. This illustrates the peak achievable T_{coeff} under different operating conditions, representing ultimate torque performance characteristic of the *SAHT*.

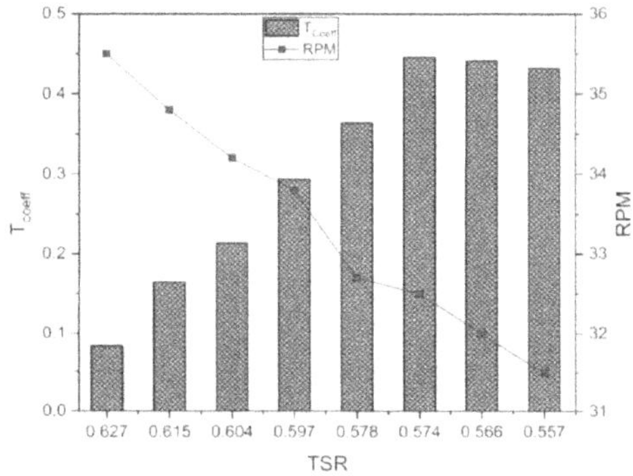

Fig. 5.43 T_{coeff} vs *TSR* with respect to turbine RPM for stream-speed of 0.8 m/s

Therefore, from the experimental study, the SAHT has shown a maximum performance till 0.8 m/s stream-speed. Furthermore, to compare the experimental works with the computational study, the CFD analysis is carried out using the best stream-speed of 0.8 m/s(as discussed in section 5.3.6 and 5.3.7).

5.7.9 Comparison of *Cp* (CFD) with *Cp* (experimental) at stream-speed 0.4m/s and 0.8 m/s

The *CFD* analysis is performed to evaluate *Cp* variation and compare with the experimental results to determine the deviation of *CFD* results for Model-5 turbine. The comparison for *SAHT* featuring an aspect ratio 0.26, gap width ratio 6.14%, and overlapping ratio 20% for stream-speed of 0.4m/s and 0.8 m/s are also determined for both CFD and experimental results as shown in Fig.5.44(a) and (b). The *CFD* results are almost similar when compared to experimental at different *TSR* values. Furthermore, both the trends in Fig.5.44(a and b) have an average error below 5% and *Cp* points and *TSR* for both *CFD* and experimental suit each other well for all *TSR* circumstances. Therefore, it can be stated that the *CFD* results are in close match with the experimental study.

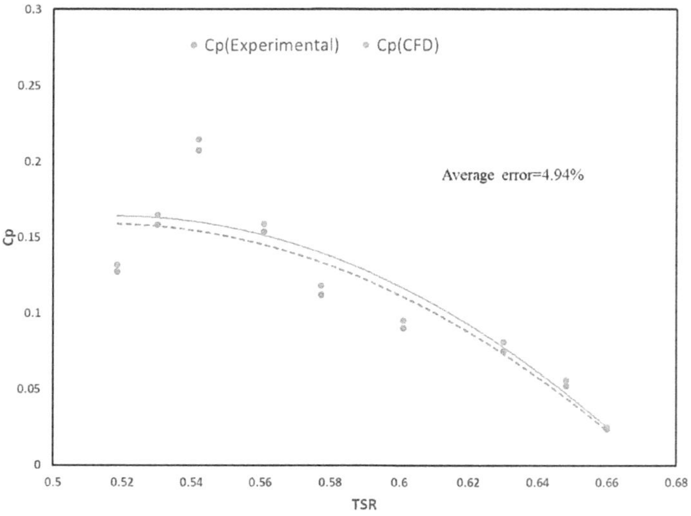

(a)

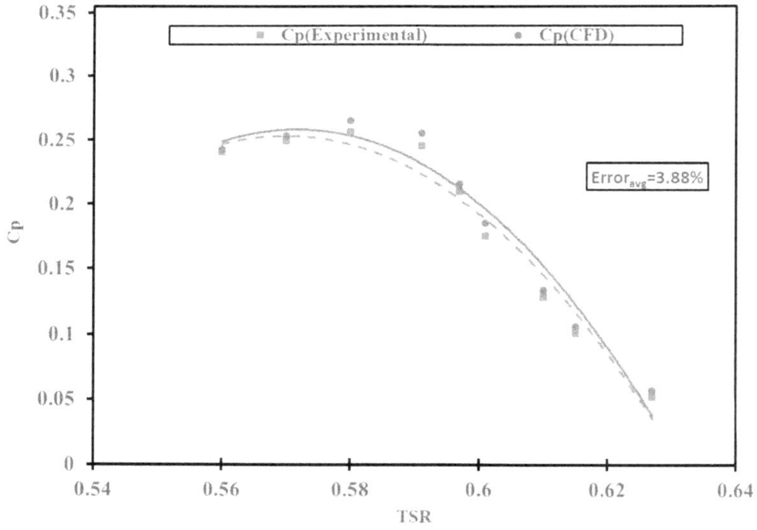

(b)

Fig.5.44 Comparison of *Cp* (CFD) with experimental study at different TSR for (a)0.4m/s and (b) 0.8m/s flow-speed

5.8 Comparison of the present study with the previous models and other published literatures

The *Cp* of the present experimental work has been contrasted with some of the published research studies where different design series of aerofoil blade profile along with the variation of gap width and overlapping ratio were implemented, as given in Table 5.7. A comparison of the current study has been performed in relation to the existing literature, focusing on similar range of low stream-speeds (0.5-1.0 m/s) and *TSR* (0.5-1). It is worth to mention that comparing based on exact input parameters from the works of different researchers is not always feasible; however, their similar levels can give meaningful comparative picture of the turbine performances. On comparing between the present design with Ayadi et al. [32], Chaudhari et al. [34], Kamal et al.[33] and Khan et al. [9] design, it is seen from Table 5.7 that both designs resemble almost the same operating conditions, whereas incorporating blade overlapping and side gap ratio, the present design can exhibit much higher *Cp* than the traditional helical Savonius water turbine. The present results indicate that utilizing a NACA 63415 blade profile with 20% overlapping ratio, and a blade side gap ratio of 6.14% achieves a maximum *Cp* of 0.256 at 0.58 *TSR* and stream-speed 0.8 m/s, which surpasses other published research studies.

Table 5.7 Results of *Cp* in relation to the published literature

Reference	Year	Blade type	Overlapping ratio, gap width ratio (%)	Stream-speed (m/s)	*TSR*	*Cp*
Khan et al. [9]	2022	S1048 airfoil	20.65 (no overlapping)	0.38	1	0.15
Kamal et al.[33]	2022	Helical with end plates	10 (no side gap ratio)	0.5	0.9	0.208
Chaudhari et al. [34]	2023	NACA 6409	15.2 (no overlapping)	1	0.8	0.227
Ayadi et al. [32]	2024	Helical Savonius	No overlap or side gap	0.86	0.7	0.18
Model-4	-	NACA 63415 airfoil-straight bladed SAHT	20,6.14	0.4	0.7	0.36
Type-3	-	NACA 63415 airfoil-straight with arc radius	20,6.14	0.5	0.7	0.278
Model-5	-	NACA 63415 airfoil	20, 6.14	0.8	0.58	0.256

5.9 Summary of the results

Two bladed Savonius hybrid hydrokinetic turbine (*SAHT*) was designed, fabricated, and investigated experimentally in the present research study. The design parameters- aspect ratio (0.22-0.29), blade side gap ratio (4.16-7.10) %, overlapping ratio (15-30) % were investigated with low stream-speed (0.6-1.0) m/s and turbine-load (200-2500) g to determine best design (Model-5) and operating conditions. Further, using *CFD* analysis the flow structures around the SAHT blades were studied and the performances of *CFD* computations were compared with the experimental results and summarized as follows:

1. Out of the three aspect ratios i.e. 0.22, 0.26, and 0.29, the *SAHT* with 0.26 performs better than the other two aspect ratios. Hence this best aspect ratio of 0.26 is recommended for the present *SAHT*.

2. The combination of blade gap width ratio (6.14%), overlapping ratio (20%) in *SAHT* produces the highest output power 0.327W and torque 0.096 Nm for 0.58 *TSR* at an operating stream-speed 0.8 m/s.

3. The maximum *Cp* (0.256) and T_{coeff} (0.445) are obtained for the stream-speed of 0.8 m/s and 0.58 *TSR*.

5.10 Experimental study of the effect of end plates on the performance of the hybrid SAHT(Model-6) model.

The best design combinations of 0.26 aspect ratio, 20% overlapping and 6.14% blade side gap ratio with addition of end plates to form a new turbine (Model-6), which is tested for a stream-speed ranging from (0.6 to 1) m/s at different load conditions (250-2000) g and results are analyzed and discussed in detail. The various hydrodynamic parameters such as Torque, *T(Nm)*, Power, *P(W)*, *Cp and* T_{coeff} are determined from the experimental results.

5.10.1 Influence of brake load over Model-6 turbine for different flow-stream speeds

Figure 5.45 illustrates the variations in *T(Nm)*, *P(W)*, *Cp*, and *TSR* as the load changes from 250 to 2000 g at a water velocity of (0.6-0.9) m/s. As the load increases, the torque steadily increases until it attains a maximum of 1500 g, and then it begins to decline as the load continues to rise beyond 1500 g. The turbine remains capable of producing work (or power) at a load of 1500 g; however, when the load increases beyond 1500 g, its efficiency begins to decline. This occurs because exceeding the 1500 g load leads to a reduction in turbine work and performance, primarily due to a decrease in turbine rpm and speed at the tip of the blades. To accommodate larger loads on the blades, the speed of the blades typically reduces, which is another cause of fall in performances. As a result, the ideal weight accommodated to the current turbine cannot be kept under 1500 g, which could lead it to operate similarly to a motor. Thus, the current turbine has produced a maximum torque of 0.096 Nm, at a stream-speed of 0.8m/s at 1500g load. Moreover, the current turbine has also shown maximum performance in terms of power (0.342W), coefficient of performance (0.268) and torque coefficient (0.446) at the same load up to 1500g and stream-speed 0.8m/s. But as the stream-speed rises above 0.8m/s, the same deteriorates. This is because the turbine blades fail to accommodate the greater thrust on the blades, which causes loss of volumetric discharge and tip speed losses at the blades resulted in reduction of performances [25]. Therefore, due to all these factors, the current turbine design performs better at 0.8m/s stream-speed.

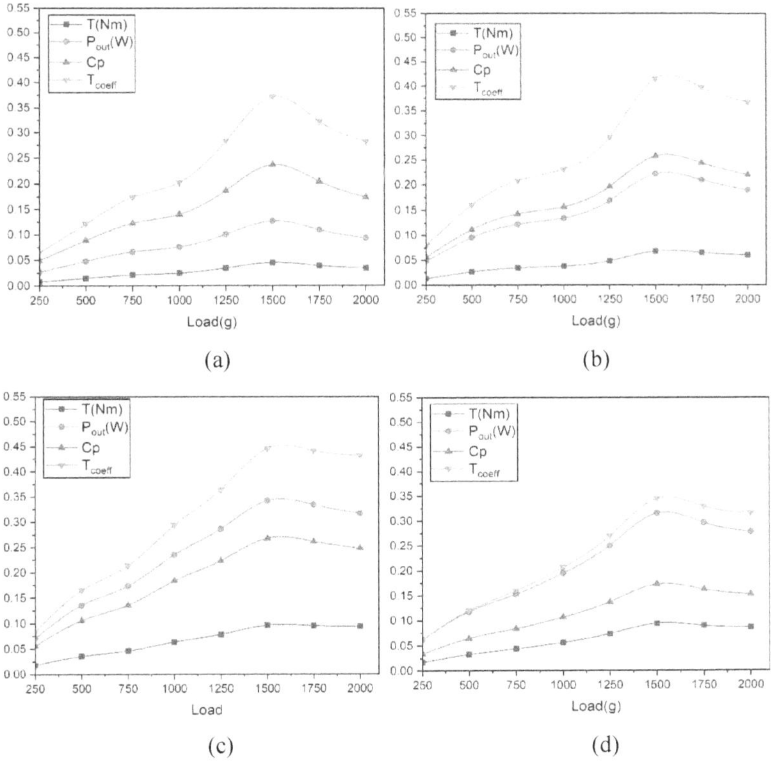

Fig.5.45 Comparison of Torque, Power, *Cp* and T_{Coeff} at different turbine loads (250-2000) g for different flow-stream speeds (a)0.6, (b)0.7m/s, (c)0.8m/s, and (d)0.9m/s

5.10.2 Influence of TSR on turbine performances at 1500g load

The effect of TSR on the turbine *T*(Nm), P_{out}(W), *Cp* and T_c and the corresponding RPM are represented in Fig.5.46(a-d). In all the cases for best stream-speed of 0.8m/s, the turbine performances rise up to 0.601 TSR and then begin to fall as the TSR decreases further. It is also observed that the RPM continuously decreases with the fall in TSR, which is due to the increases of load at the turbine. But the current turbine can still perform better at 0.601 TSR though the RPM is low (i.e.34 RPM) at 0.601 TSR, which is because, the turbine can accommodate maximum load of 1500g at 0.601 TSR and produces highest torque, power, *Cp* and T_{coeff} at 0.8m/s stream-speed as seen from Fig.5.46(a-d). But as the load rises above 1500g at same stream-speed, the TSR not only decreases but also the performances reduce with the fall in RPM and the lowest RPM obtained at 0.574 TSR is 32.5 RPM. Thus from the analysis, it can be stated that the addition of end plate has increased the overall performances of the turbine up to higher TSR of 0.601

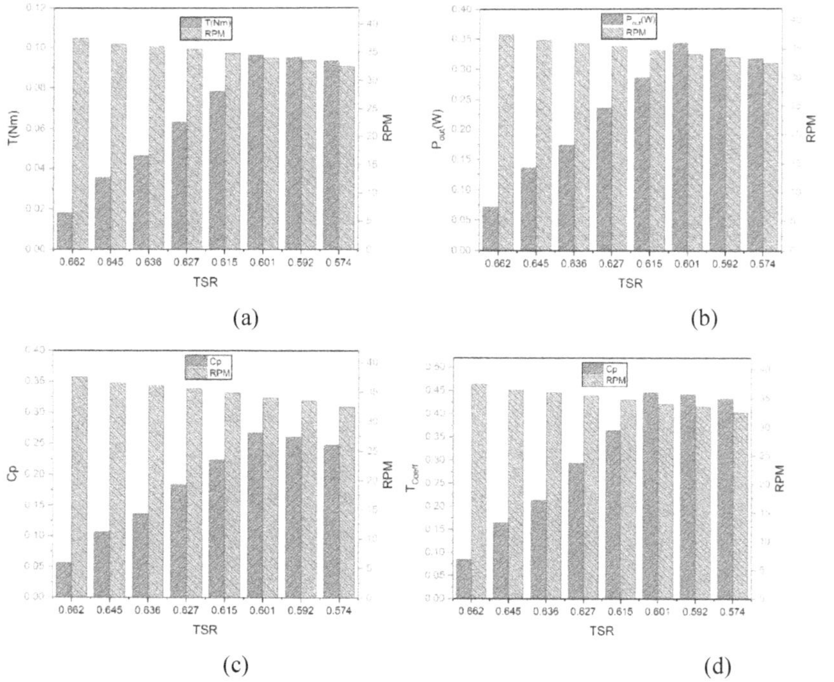

Fig.5.46 Comparison of Torque(Nm), $P_{out}(W)$, *Cp* and T_c with respect to *TSR* at flow-stream speeds 0.8 m/s

5.10.3 Comparison of maximum *Cp* and T_{Coeff} for various flow-stream speeds

Figure 5.47(a) shows the influence of *TSR* over Max. *Cp* on *SAHT* at different stream-speeds (0.6-1) m/s for optimal gap width 6.14% and 20% overlapping at 1500g load. It exhibits the maximum *Cp* that may be attained under different operating conditions, signifying the SHKT's final performance characteristics. In Fig.5.47(a), the *Cp* rises to maximum of 0.268 at 0.601 *TSR*, which is obtained at 0.8m/s stream-speed and then it decreases as the stream-speed increases above 0.8m/s. The lowest *Cp* obtained is 0.116 at 0.431 *TSR* for 1m/s stream-speed. Similarly, for Fig.5.47(b), the max T_{coeff} of 0.446 is obtained at 0.601 TSR and then falls as TSR rises further. This is because the turbine blades' size, projected area, blade side gap, and overlapping enable water to exert maximum desirable force over blades by accepting maximum stream-speed up to 0.8 m/s at 0.601 *TSR* for 1500 g load. However, the stream-speed above 0.8m/s, causes tip loss and the water leaves turbine without affecting the blades, which reduces performances of the turbine.

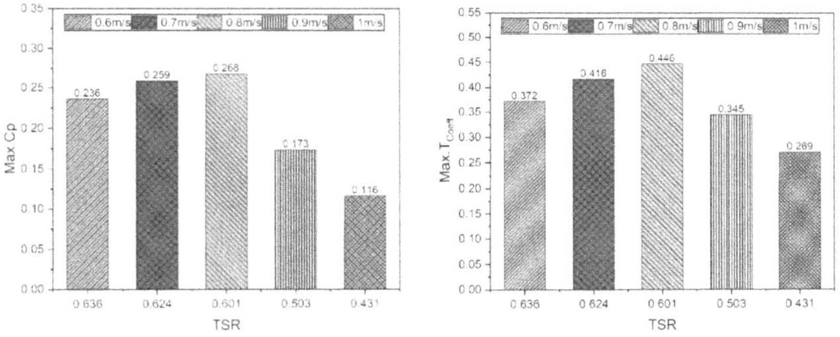

(a) (b)

Fig.5.47 Effect of *TSR* on (a)Max. *Cp* and (b)Max. T_{coeff} for *SAHT* at a stream-speed of (0.6-1.0) m/s

5.10.4 Comparison with the previous models and other published literatures

The *Cp* of the present experimental work has been contrasted with some of the published research studies where different design series of aerofoil blade profile along with the variation of gap width and overlapping ratio were implemented, as given in Table 5.7. A comparison of the current study has been performed in relation to the existing literature, focusing on similar range of low stream-speeds (0.5-1.0 m/s) and *TSR* (0.5-1). It is worth to mention that comparing based on exact input parameters from the works of different researchers is not always feasible; however, their similar levels can give meaningful comparative picture of the turbine performances. On comparing between the present design with Ayadi et al. [32], Chaudhari et al. [34], Kamal et al.[33] and Khan et al. [9] design, it is seen from Table 5.7 that both designs resemble almost the same operating conditions, whereas incorporating blade overlapping and side gap ratio, the present design can exhibit much higher *Cp* than the traditional helical Savonius water turbine. The present results indicate that utilizing a NACA 63415 blade profile with 20% overlapping ratio, and a blade side gap ratio of 6.14% and end plates achieves a maximum *Cp* of 0.268 at 0.6 *TSR* and stream-speed 0.8 m/s, which surpasses other published research studies.

Table 5.8 Comparison of Cp with the previous models and published literatures

Reference	Year	Blade type	Type of Study	Overlapping ratio, gap width ratio (%)	Stream-speed (m/s)	*TSR*	*Cp*
Khan et al. [9]	2022	S1048 airfoil	CFD	20.65 (no overlapping)	0.38	1	0.15
Kamal et al.[33]	2022	Helical with end plates	CFD	10 (no side gap ratio)	0.5	0.9	0.208
Chaudhari et al. [34]	2023	NACA 6409	CFD	15.2 (no overlapping)	1	0.8	0.227
Ayadi et al.[32]	2024	Helical Savonius	CFD	No overlap or side gap	0.86	0.7	0.18
Khani et al.[35]	2023	Semi-circular two bladed Savonius turbine	CFD	No overlap or side gap	0.55	0.89	0.28
Model-4	-	**NACA 63415 airfoil-straight bladed SAHT**	CFD	**20,6.14**	**0.4**	**0.7**	**0.36**
Type-2	-	NACA 63415 airfoil-straight with arc radius	CFD	20,6.14	0.5	0.7	0.278
Model-5	-	NACA 63415 airfoil-straight bladed SAHT	Experimental	20, 6.14	0.8	0.58	0.256
Model-6	-	NACA 63415 airfoil-straight bladed SAHT with end plates	Experimental	20,6.14	0.8	0.6	0.268

5.11 Summary of the results

In this work, Model-5 was fabricated with the addition of end plates on the top and bottom ends of the turbine to form Model-6 and experimentally investigated in an open water flow channel under (200-2500) g turbine loads and low flow-speeds ≤1m/s to improve its performances. Further the *Cp* of Model-6 is compared with the previous SAHT models and other published literatures. The summary of the experimental outcomes are as follows:

- The current turbine has produced a maximum torque of 0.096 Nm, at a stream-speed of 0.8m/s at 1500g load.
- Moreover, the current turbine has also shown maximum performance in terms of power(0.342W), torque coefficient (0.446) at the same load up to 1500g and stream-speed 0.8m/s.
- The maximum power coefficient of 0.268 at TSR 0.601 was obtained at flow speed 0.8 m/s and brake load 1500 g for A_r =0.26 with end plates and with optimal blade design parameters.
- The addition of end plate has increased the overall performances of the turbine up to higher TSR of 0.601.
- Experiments revealed that partial immersion of the blades due to considering different aspect ratios did not improve the turbine performance. Rather putting end plates on the top and bottom ends of the turbine improved its performance.

References

[1] Patel, V., Bhat, G., Eldho, T.I. and Prabhu, S.V., 2017. Influence of overlap ratio and aspect ratio on the performance of Savonius hydrokinetic turbine. *International Journal of Energy Research*, *41*(6), pp.829-844.

[2] Frikha, S., Driss, Z., Ayadi, E., Masmoudi, Z. and Abid, M.S., 2016. Numerical and experimental characterization of multi-stage Savonius rotors. *Energy*, *114*, pp.382-404.

[3] Sarma, K.C., Biswas, A., Nath, B. and Misra, R.D., 2024. Effect of overlapping and space between stages of a three-bladed double-stage Savonius hydro-kinetic turbine for low flow-speed perennial river application. *Energy Sources, Part A: Recovery, Utilization, and Environmental Effects, 46(1)*, pp.1177-1195.

[4] Yap, T.C., Abdullah, M.Z., Husain, Z., Ripin, Z.M. and Ahmad, R., 2001,

December. The effect of turbulence intensity on the aerodynamic performance of airfoils. In *4th International Conference on Mechanical Engineering* (pp. 31-36).

[5] Hoffmann, J.A., 1991. Effects of freestream turbulence on the performance characteristics of an airfoil. *AIAA journal, 29*(9), pp.1353-1354.

[6] Karasu, İ. and Tümse, S., 2021. The turbulence intensity effect on the flow characteristics and aerodynamics of a circular cylinder. Çukurova Üniversitesi Mühendislik Fakültesi Dergisi, 36(4), pp.901-912.

[7] Zhang, Y.N., Zhang, M.M., Cai, C. and Xu, J.Z., 2020. Aerodynamic load control on a dynamically pitching wind turbine airfoil using leading-edge protuberance method. *Acta Mechanica Sinica*, 36, pp.275-289.

[8] Lee, J.H., Lee, Y.T. and Lim, H.C., 2016. Effect of twist angle on the performance of Savonius wind turbine. *Renewable Energy*, *89*, pp.231-244.

[9] Khan, Z.U., Ali, Z. and Uddin, E., 2022. Performance enhancement of vertical axis hydrokinetic turbine using novel blade profile. *Renewable Energy*, *188*, pp.801-818.

[10] Biber, K. and Zumwalt, G.W., 1993. Hysteresis effects on wind tunnel measurements of a two-element airfoil. *AIAA journal, 31(2)*, pp.326-330.

[11] Celik, Y., Ingham, D., Ma, L. and Pourkashanian, M., 2023. Novel hybrid blade design and its impact on the overall and self-starting performance of a three-dimensional H-type Darrieus wind turbine. *Journal of Fluids and Structures, 119*, p.103876.

[12] Hu, J., Wang, Z., Chen, C. and Guo, C., 2020. Vortex shedding simulation of hydrofoils with trailing-edge truncation. Ocean Engineering, 214, p.107529.

[13] Launder, B.E. and Spalding, D.B., 1983. The numerical computation of turbulent flows. Numerical Prediction of Flow. *Heat Transfer, Turbulence and Combustion*, *3*, pp.96-116.

[14] Chakraborty, P., Balachandar, S. and Adrian, R.J., 2007. Kinematics of local vortex identification criteria. *Journal of visualization*, *10*(2), pp.137-140.

[15] Dubief, Y. and Delcayre, F., 2000. On coherent-vortex identification in turbulence. *Journal of turbulence*, *1*(1), p.011.

[16] Basumatary, M., Biswas, A. and Misra, R.D., 2021. Experimental verification of improved performance of Savonius turbine with a combined lift and drag based blade profile for ultra-low head river application. *Sustainable Energy Technologies and Assessments*, *44*, p.100999.

[17] Shashikumar, C.M., Vijaykumar, H. and Vasudeva, M., 2021. Numerical investigation of conventional and tapered Savonius hydrokinetic turbines for low-velocity hydropower application in an irrigation channel. Sustainable Energy Technologies and Assessments, 43, p.100871.

[18] Nath, B., Biswas, A., Das, B. and Misra, R.D., 2024. Design, optimization and analysis of a modified Savonius hydro-kinetic turbine (MSHT) with curved winglet and straight blade. *Energy Conversion and Management, 314*, p.118699.

[19] Rogowski, K., Królak, G. and Bangga, G., 2021. Numerical study on the aerodynamic characteristics of the NACA 0018 airfoil at low Reynolds number for Darrieus wind turbines using the transition SST model. *Processes, 9(3)*, p.477.

[20] Choudhry, A., Arjomandi, M. and Kelso, R., 2015. A study of long separation bubble on thick airfoils and its consequent effects. *International Journal of Heat and Fluid Flow*, 52, pp.84-96.

[21] Mohan, M., Kansagara, D.D., Sharma, D. and Saha, U.K., 2021, December. Estimating the Aerodynamic Coefficients of a Savonius Rotor Blade Profile Developed Through the Simplex Search Method. In *Gas Turbine India Conference* (Vol. 85536, p. V001T09A003). American Society of Mechanical Engineers.

[22] Homann, H., Bec, J. and Grauer, R., 2013. Effect of turbulent fluctuations on the drag and lift forces on a towed sphere and its boundary layer. *Journal of Fluid Mechanics*, *721*, pp.155-179.

[23] Kadam, A.A. and Patil, S.S., 2013. A review study on Savonius wind rotors for accessing the power performance. *IOSR Journal of Mechanical and Civil Engineering*, *5*, pp.18-24.

[24] Sarma, K.C., Nath, B., Biswas, A. and Misra, R.D., 2023. Design and performance investigation of a triple blade dual stage Savonius-alike hydrokinetic turbine from low flow stream reserves. *Energy Sources, Part A: Recovery, Utilization, and Environmental Effects*, *45*(4), pp.12099-12117.

[25] Sarma, K.C., Biswas, A., Nath, B. and Misra, R.D., 2024. Experimental study of the effect of gap between the stages of a double-stage Savonius Hydrokinetic turbine with end plates and blade overlapping at low water currents. *Energy Sources, Part A: Recovery, Utilization, and Environmental Effects*, *46*(1), pp.10557-10578.

[26] Basumatary, M., Biswas, A. and Misra, R.D., 2018. CFD analysis of an innovative combined lift and drag (CLD) based modified Savonius water turbine. *Energy Conversion and Management*, *174*, pp.72-87.

[27] Yigit, C., 2024. Numerical investigation of specific performance parameters of the S-ROTOR; a Savonius type turbine design. *Ocean Engineering*, *291*, p.116314.

[28] Hassanzadeh, R. and Mohammadnejad, M., 2019. Effects of inward and outward overlap ratios on the two-blade Savonius type of vertical axis wind turbine performance. *International Journal of Green Energy*, *16*(15), pp.1485-1496.

[29] Babu, G.V. and Patel, D.K., 2024. Performance analysis of a Savonius hydrokinetic turbine with aerodynamic blade shape and different overlap ratios: A numerical flow-driven approach. *Ocean Engineering*, *304*, p.117868.

[30] Thiyagaraj, J., Rahamathullah, I., Anbuchezhiyan, G., Barathiraja, R. and Ponshanmugakumar, A., 2021. Influence of blade numbers, overlap ratio and modified blades on performance characteristics of the Savonius hydro-kinetic turbine. *Materials Today: Proceedings*, *46*, pp.4047-4053.

[31] Menter, F.R., 1994. Two-equation eddy-viscosity turbulence models for engineering applications. AIAA journal, 32(8), pp.1598-1605.

[32] Ayadi, A., Rabeh, N., Nasraoui, H., Mosbahi, M. and Driss, Z., Enhancing the efficiency of hydrokinetic Savonius turbine with guiding walls. *Environmental Progress & Sustainable Energy*, p.e14532.

[33] Kamal, M.M. and Saini, R.P., 2022. A numerical investigation on the influence of Savonius blade helicity on the performance characteristics of hybrid cross-flow hydro-kinetic turbine. *Renewable Energy*, *190*, pp.788-804.

[34] Chaudhari, V.N. and Shah, S.P., 2023. Numerical investigation on the performance of an innovative Airfoil-Bladed Savonius Hydro-kinetic Turbine (ABSHKT) with deflector. International Journal of Thermofluids, p.100279.

[35] Khani, M.S., Shahsavani, Y., Mehraein, M. and Kisi, O., 2023. Performance evaluation of the savonius hydrokinetic turbine using soft computing techniques. *Renewable Energy*, *215*, p.118906.

CHAPTER 6

CONCLUSIONS AND FUTURE SCOPE

6.1 Conclusions

Water has been a source of energy for many decades. The utility of water as a source of energy has been extended further to drive water turbines thereby producing power. In this work, different model designs of hybrid SAHTs were computationally and experimentally investigated in detail in a recirculating water flow channel under different design conditions and different low flow speeds $\leq$1m/s, commonly seen in perennial river, channel, canal, etc., in order to improve SAHT performance under such flow speed conditions and low TSR conditions. The conclusions of this study are summarized as follows:

- An efficient combined lift and drag based modified zero-head Savonius hydro-kinetic turbine has been designed and developed for Pico-scale power generation from low flow speeds using computational and experimental approaches.
- Different prominent lift-based symmetrical/unsymmetrical airfoil profiles from NACA and S-series airfoils were investigated in terms of lift-to drag ratio to select the best airfoil profile for integrating in the turbine blade along with drag-based straight blade profile to make it a combined lift and drag based modified Savonius turbine.
- It was found that NACA series unsymmetrical profiles have better hydrodynamic performance than S-series profiles. And out of considered NACA series blades, NACA 63415 airfoil was selected as it possessed the highest lift-to-drag ratio 52.7.
- The study shows that innovative design of the blade profile with a new concept of offset between the cut-out arc section of airfoil and straight blade section, then proper cut-out arc section for the airfoil can advance the performance apart from more than by only taking the route of conventional parameters like

Overlapping ratio or side clearance ratio for the development of Savonius hydrokinetic turbine.

- Further, optimization of these blade parameters for improved performance of the turbine is also important, as optimal overlapping of 20%, side clearance of 6.14%, offset length of 15 mm, and also cut-out arc section of 12% demonstrated the highest power coefficient and torque coefficient of the hybrid SAHT Savonius turbine.
- Flow structure across the developed turbine reflected the positive hydrodynamics performances of the selected design parameters, leading to effective blade-fluid flow control-based interactions. Velocity vectors from offset section of the upstream blade improved the velocity signature on suction surface of the airfoil cut out arc section. Overlapping and side clearance flow strengthened the flow thrust on the straight blade section. Furthermore, Coanda-like vectors from downstream also contributed to flow thrust on the blades.
- Best Cut-out arc section of 12% contributed to higher performance than the hybrid SAHT turbine with optimal overlapping, side clearance, and offset. Flow structure (Q-criterion) revealed that vortex streak separating from the suction surface of curved section of upstream-side blade was much subdued and less vortices were prevalent across the straight section and the overlapping zone for the best cut-out arc section case.
- The highest coefficient of performance (Cp), and torque coefficient (T_{coeff}) of 0.36 and 0.75, respectively are obtained for Model-4 with 12% (best cut out arc section) and winglet arc angle (a1) 22.7° along with the Taguchi optimized design combinations of 20 % blade overlapping, 6.14 % gap width ratio, and 15 mm winglet offset at low flow speed 0.4 m/s and TSR 0.7.
- Among the variants of Model 4 (i.e. Type 1 to Type 3), the results show that winglet profile with 15 mm arc curvature radius (Type 2) along with 20% blade overlapping, 6.14 % gap width ratio, and 15 mm winglet offset exhibits the highest Cp of 0.278 and T_{coeff} of 0.702, respectively at TSR 0.7.
- Experiments revealed that partial immersion of the blades due to considering different aspect ratios did not improve the turbine performance. Rather putting end plates on the top and bottom ends of the turbine improved its performance.

The maximum power coefficient of 0.268 at TSR 0.601 was obtained at flow speed 0.8 m/s and brake load 1500 g for A_R=0.26 with end plates and with optimal blade design parameters.

- CFD analysis of the experimental model with reduced diameter reveals similar performance as experiments within an error of 4.94% for flow speed 0.4 m/s and within an error of 3.88% for flow speed 0.8 m/s.
- However, among all the experimental and CFD models, the highest performance (Cp = 0.36) is obtained at 0.4 m/s flow speed for the overall turbine diameter of 334.63 mm for Model-4 and same turbine height as experiment (i.e. 70 mm) with the optimal blade design parameters.

6.2 Scope for future works

In order to upgrade the present state-of-art in the area of water turbine for pico-scale applications and based on the outcome of the present work, the following works are identified as the scope for future works:

Short-Term Research Priorities

1. Optimization of the SAHT blade geometry and operating parameters to enhance performance under ultra-low flow velocities below **0.3 m/s**.
2. CFD-based investigation of SAHT configurations with an increased number of blades to assess their influence on torque generation, starting characteristics, and overall efficiency.
3. Appropriate location of blending of curved section with straight section may reduce flow separation and recirculation.
4. 3D simulation with end plates for the optimized SAHT model could be performed for identifying the blade tip effects.

Long-Term Research Priorities

1. Development and performance evaluation of multi-stage SAHT configurations operating under increased flow depths to improve energy extraction capability.
2. Integration of flow-guiding devices, such as deflectors and curtain plates, to control incoming flow structures and further enhance turbine performance under varying hydrodynamic conditions.

LIST OF PUBLICATIONS

International Journals

1) **Nath, B.,** Biswas, A., Das, B. and Misra, R.D., 2024. Design, optimization and analysis of a modified Savonius hydro-kinetic turbine (HYBRID SAHT) with curved winglet and straight blade. *Energy Conversion and Management*, *314*, p.118699. https://doi.org/10.1016/j.enconman.2024.118699
2) **Nath, B.,** Biswas, A., Das, B. and Misra, R.D., 2024. Hydrodynamic Investigation of a Hybrid-Bladed Savonius Hydrokinetic Turbine-Computational Fluid Dynamics and Taguchi Optimization Approaches. *Journal of Fluids Engineering*, pp.1-31. https://doi.org/10.1115/1.4068402
3) **Nath, B.,** Biswas, A., Das, B. and Misra, R.D., 2025. Design and flow analysis of a Savonius-type hydro-kinetic turbine with hybrid-blade profile. *Environmental Progress & Sustainable Energy*, p.e70173.
4) **Nath, B.,** Biswas, A., Das, B. and Misra, R.D., 2026. Experimental and fluid-flow structure analysis of a Savonius-resembling hybrid water turbine at low stream-speeds. *Journal of the Brazilian Society of Mechanical Sciences and Engineering*, *48*(2), p.99.

Conference

Nath, B., Biswas, A., Das, B. and Misra, R.D., 2025. Flow Structure Analysis of Hybrid Hydrokinetic Savonius Turbine with combination of Straight and Curve Blade – A Computational Study. In *Recent Advancements in Mechanical Engineering: Select Proceedings of ICRAME 2025.*

Patent

Title: A Savonius Hydro-Kinetic Turbine Assembly. **Indian Patent Application Number- 202531013159(2025)**. Name of the Inventors- **Biswajit Nath**, Agnimitra Biswas, Biplab Das, Rahul Dev Misra. **Status-**Published.

CONTENTS